Advances of Accurate Quantification Methods in Food Analysis

Advances of Accurate Quantification Methods in Food Analysis

Editors

Xianjiang Li
Rui Weng

MDPI • Basel • Beijing • Wuhan • Barcelona • Belgrade • Manchester • Tokyo • Cluj • Tianjin

Editors
Xianjiang Li
National Institute of
Metrology China
China

Rui Weng
Chinese Academy of
Agricultural Sciences
China

Editorial Office
MDPI
St. Alban-Anlage 66
4052 Basel, Switzerland

This is a reprint of articles from the Special Issue published online in the open access journal *Separations* (ISSN 2297-8739) (available at: https://www.mdpi.com/journal/separations/special_issues/Accurate_Quantification).

For citation purposes, cite each article independently as indicated on the article page online and as indicated below:

LastName, A.A.; LastName, B.B.; LastName, C.C. Article Title. *Journal Name* **Year**, *Volume Number*, Page Range.

ISBN 978-3-0365-5843-1 (Hbk)
ISBN 978-3-0365-5844-8 (PDF)

Cover image courtesy of Xianjiang Li

© 2022 by the authors. Articles in this book are Open Access and distributed under the Creative Commons Attribution (CC BY) license, which allows users to download, copy and build upon published articles, as long as the author and publisher are properly credited, which ensures maximum dissemination and a wider impact of our publications.

The book as a whole is distributed by MDPI under the terms and conditions of the Creative Commons license CC BY-NC-ND.

Contents

About the Editors vii

Preface to "Advances of Accurate Quantification Methods in Food Analysis" ix

Xianjiang Li and Rui Weng
Advances in Accurate Quantification Methods in Food Analysis
Reprinted from: *Separations* 2022, 9, 342, doi:10.3390/separations9110342 1

Xianjiang Li, Sheng Wang, Zhen Guo, Xiuqin Li, Qinghe Zhang and Hongmei Li
Determination of Fosetyl-Aluminum in Wheat Flour with Extract-Dilute-Shoot Procedure and Hydrophilic Interaction Liquid Chromatography Tandem Mass Spectrometry
Reprinted from: *Separations* 2021, 8, 197, doi:10.3390/separations8110197 5

Kaixuan Tong, Yujie Xie, Siqi Huang, Yongcheng Liu, Xingqiang Wu, Chunlin Fan, Hui Chen, Meiling Lu and Wenwen Wang
QuEChERS Method Combined with Gas- and Liquid-Chromatography High Resolution Mass Spectrometry to Screen and Confirm 237 Pesticides and Metabolites in Cottonseed Hull
Reprinted from: *Separations* 2022, 9, 91, doi:10.3390/separations9040091 15

Xingqiang Wu, Kaixuan Tong, Changyou Yu, Shuang Hou, Yujie Xie, Chunlin Fan, Hui Chen, Meiling Lu and Wenwen Wang
Development of a High-Throughput Screening Analysis for 195 Pesticides in Raw Milk by Modified QuEChERS Sample Preparation and Liquid Chromatography Quadrupole Time-of-Flight Mass Spectrometry
Reprinted from: *Separations* 2022, 9, 98, doi:10.3390/separations9040098 35

Rui Bie, Jiguang Zhang, Yunbai Wang, Dongmei Jin, Rui Yin, Bin Jiang and Jianmin Cao
Analysis of Multiclass Pesticide Residues in Tobacco by Gas Chromatography Quadrupole Time-of-Flight Mass Spectrometry Combined with Mini Solid-Phase Extraction
Reprinted from: *Separations* 2022, 9, 104, doi:10.3390/separations9050104 53

Youyou Yang, Zhuolin He, Lei Mu, Yunfeng Xie and Liang Wang
Simultaneous Determination of 23 Mycotoxins in Broiler Tissues by Solid Phase Extraction UHPLC-Q/Orbitrap High Resolution Mass Spectrometry
Reprinted from: *Separations* 2021, 8, 236, doi:10.3390/separations8120236 67

Yueting Xiao, Shuyu Liu, Yuan Gao, Yan Zhang, Qinghe Zhang and Xiuqin Li
Determination of Antibiotic Residues in Aquaculture Products by Liquid Chromatography Tandem Mass Spectrometry: Recent Trends and Developments from 2010 to 2020
Reprinted from: *Separations* 2022, 9, 35, doi:10.3390/separations9020035 81

Xiao Li, Qian Ma, Chao Wei, Wei Cai, Huanhuan Chen, Rui Xing and Panshu Song
Green and Simple Extraction of Arsenic Species from Rice Flour Using a Novel Ultrasound-Assisted Enzymatic Hydrolysis Method
Reprinted from: *Separations* 2022, 9, 105, doi:10.3390/separations9050105 97

Yue Zhao, Min Wang, Mengrui Yang, Jian Zhou and Tongtong Wang
Determination of Selenomethionine, Selenocystine, and Methylselenocysteine in Egg Sample by High Performance Liquid Chromatography—Inductively Coupled Plasma Mass Spectrometry
Reprinted from: *Separations* 2022, 9, 21, doi:10.3390/separations9020021 107

Xue Li, Wei Wang, Suling Sun, Junhong Wang, Jiahong Zhu, Feng Liang, Yu Zhang and Guixian Hu
Quantitative Analysis of Anthocyanins in Grapes by UPLC-Q-TOF MS Combined with QAMS
Reprinted from: *Separations* **2022**, *9*, 140, doi:10.3390/separations9060140 . . . . **121**

About the Editors

Xianjiang Li

Dr. Xianjiang Li received his Ph.D. degree from Peking University then worked as a post-doc researcher at National Institute of Metrology (NIM). Subsequently, he worked as a visiting scientist at Bureau International des Poids et Mesures (BIPM). He is a Level 1 certified metrology engineer, as well as being a technical expert in CRM certification. Currently, he works as an associate research fellow at NIM. His current research interests include high-accuracy method development for food safety and the development of certified reference materials. He has chaired one project for the National Natural Science Foundation of China, as well as participating in three projects of the National Key R&D Program of China. He has published 49 papers with more than 1600 citations (h-index = 22), including 4 on *Food Chem.*, 2 on *Coord. Chem. Rev.* and 1 on *Trends Food Sci. Technol.* He has also chapters in two English books and one Chinese book. He also works as editor for *J. Anal. Test.*, guest editor for Separations and review editor for *Front. Nutrition*. He is invited as reviewer for 27 journals, including for *Food Chem., Food Control, J. Hazard. Mater., Environ., Sci. Nano*, and so on.

Rui Weng

Rui Weng (Dr.) received her Ph.D. degree from Peking University and now works as an associate researcher at the Institute of Quality Standard and Testing Technology for Agro-Products of the Chinese Academy of Agricultural Sciences. Her major research interest is non-targeted screening of pesticides and veterinary drugs, and the quality evaluation of agro-products. She has chaired 12 projects from the National Natural Science Foundation of China, the National Key R&D Program of China, the Beijing Science and Technology Program, etc. She has received five national invention patents and two software copyrights, publishing 32 papers in journals such as *Nano Today, Anal. Chem., Food Chem.*, among others. She also works as a reviewer for *Food Chem., Anal. Bioanal. Chem.*, and so on.

Preface to "Advances of Accurate Quantification Methods in Food Analysis"

The nine articles published in this Special Issue cover the latest advances in food analysis methods. On the one hand, a suitable pretreatment method is the key to eliminating matrix effects and lowering the detection limit, which arises due to the complexity of food components; on the other hand, HRMS exhibits an excellent performance, with a trend towards multi-residue and multi-class detection. We strongly encourage the publication of further Special Issues in this field to closely monitor current research.

Xianjiang Li and Rui Weng
Editors

Editorial

Advances in Accurate Quantification Methods in Food Analysis

Xianjiang Li [1,*] and Rui Weng [2,*]

[1] Key Laboratory of Chemical Metrology and Applications on Nutrition and Health for State Market Regulation, Division of Metrology in Chemistry, National Institute of Metrology, Beijing 100029, China
[2] Key Laboratory of Agro-Food Safety and Quality of Ministry of Agriculture and Rural Affairs, Institute of Quality Standard and Testing Technology for Agro-Products, Chinese Academy of Agricultural Sciences, Beijing 100081, China
* Correspondence: lixianjiang@nim.ac.cn (X.L.); wengrui@caas.ac.cn (R.W.)

1. Introduction

Food safety is an important topic, and with the perfection of regulations and technologies, food safety is improving. However, incidents such as "toxic mineral oil" [1] and the "fipronil egg scandal" [2] occasionally trigger public tension. In addition to hazardous analytes, increasing attention has been paid to food nutrition with the aim of improving people's lives. Therefore, many researchers are working to develop and validate analytical methods to identify and quantify hazardous and nutritional analytes in foods.

In recent years, there has been especially rapid growth in the accurate quantification of food components such as pesticides [3–5], veterinary drugs [6], mycotoxins [7], amino acids [8], nucleotides [9] and organosulfur compounds [10]. Concerning food analysis, sample pretreatment aims to enrich target analytes and remove complicated matrix components (lipids, proteins, salts, acids, pigments, etc.). New extraction and purification strategies provide high specificity and efficiency for the targets. Combined strategies exhibit excellent performance in lowering matrix effects. Following this, gas/liquid chromatography–tandem mass spectrometry (GC/LC–MS/MS) is a powerful tool in guaranteeing food safety and quality. Moreover, the use of high-resolution mass spectrometry (HRMS) allows the high-throughput identification and screening of targeted/untargeted analytes.

This Special Issue welcomes any developments in novel sample pretreatment or detection techniques to realize accurate quantification in food analysis.

2. Summary of Published Articles

This Special Issue includes nine manuscripts which address the latest analytical methods for the identification and characterization of a variety of hazardous and nutritional compounds in foods.

We received four manuscripts on the topic of pesticide determination. The first manuscript is from guest editor Dr. Li, who reported an LC–MS/MS method for single-pesticide analysis in wheat flour [11]. Fosetyl-aluminum is an ionic fungicide; therefore, an extract–dilute–shoot strategy was developed with water and acetonitrile. Then, fosetyl was quantified via hydrophilic interaction liquid chromatography–tandem mass spectrometry. Finally, the developed method was used to analyze 75 wheat flour samples collected from four provinces in China. The other three manuscripts focus on multi-pesticide analysis via HRMS. Chen and coworkers coupled the QuEChERS method with time-of-flight mass spectrometry for hundreds of pesticides in cottonseed hull [12] and raw milk samples [13]. For the oily samples, acetonitrile with 1% acetic acid was selected as an extraction solvent, followed by purification with $MgSO_4$, C_{18}, or primary secondary amine sorbents. Satisfactory recovery was achieved with three spiking levels. To simplify the sample pretreatment procedure, Cao's group used the min-SPE strategy to replace QuEChERS in an analysis of 209 pesticides in tobacco [14]. The Box–Behnken design was used to optimize the parameters of water, solvent and purification volume. The commercial min-SPE device

demonstrated good performance in the removal of typical tobacco matrix interferences, including nicotine, nicotyrine and anatabine. A database library was homebuilt for qualitative and quantitative workflow, with a default matching score ≥ 75. For real sample analysis, five positive samples were quantified with matrix-matched standards.

The fifth article relates to mycotoxin detection in broiler tissues using Orbitrap [15]. An acetonitrile–water–formic acid mixture was used as an extraction solvent. A database was built with information on retention time and accurate mass for the qualitative screening and simultaneous quantification of 23 mycotoxins. Using this database, data were acquired via the full scan and data-dependent MS/MS modes. The extra-high resolution (70,000 for MS^1 and 17,500 for MS^2) was effective at avoiding false-positive results. Finally, one chicken liver was determined to be positive for zearalenone among 30 collected samples.

The sixth manuscript is a review of antibiotic determination in aquatic products from 2010 to 2020 [16]. Li and coworkers summarized the typical sample pretreatment techniques, including liquid–liquid extraction, solid-phase extraction, QuEChERS, pressurized liquid-phase extraction and microwave-assisted extraction. Since some antibiotics are bound to proteins, proper hydrolysis is required to free them and achieve accurate results. Matrix effects are common in ionization processes; thus, elimination or compensation is very important for reliable results. The authors summarized the available matrix certified reference materials from China, Australia, South Korea and Canada, which are important for the traceability of chemical measurement.

We also received one article describing the development of a method for hazardous metal ions. The work focused on arsenic (As) species, and the rice flour certified reference material NMIJ-7532a was used for method development [17]. Herein, ultrasound-assisted enzymatic hydrolysis with α-amylase was adopted to liberate As(III), As(V), monomethylarsonate, dimethylarsinic acid, arsenobetaine and arsenocholine. Then, the As species were quantified using an inductively coupled plasma mass spectrometry (ICP-MS) method. Based on the assigned values, the recovery of As(III), As(V) and dimethylarsinic acid was better than 98%.

The last two manuscripts are about nutritional analytes. Wang's group described an analysis method for three organic selenium (Se) species (selenomethionine, selenocystine and methylselenocysteine) in egg samples [18]. Since Se was widely incorporated into protein, enzymatic hydrolysis with protease XIV was optimized in sample preparation. Moreover, lipase was added to remove fat interference. Then, three targeted species were separated using ion-pairing reversed-phase chromatography systems and determined using ICP-MS. Real sample analysis demonstrated that Se-enriched eggs had more total Se and selenomethionine, while there was no significant difference in the contents of selenocystine and methylselenocysteine. Anthocyanins are complex natural compounds with limited and costly standards. Therefore, Hu and coworkers analyzed them via quantitative analysis of multi-components using single-marker QSAM with high-resolution MS [19]. Peonidin 3-O-glucoside was used as an internal reference for the nine anthocyanins. The relative correction factor was calculated using the multi-point and slope methods, and the slope method was more accurate. Additionally, there were no significant deviations in the anthocyanin content between QSAM and external standard method. Satisfactory results were obtained with the real samples.

3. Conclusions

The nine articles published in this Special Issue cover the latest advances in food analysis methods. On the one hand, a suitable pretreatment method is the key to eliminating matrix effects and lowering the detection limit, which arises due to the complexity of food components; on the other hand, HRMS exhibits excellent performance, with a trend towards multi-residue and multi-class detection. We highly encourage the creation of more Special Issues in this field to closely track current research.

Author Contributions: Conceptualization and writing, X.L.; review and editing, R.W. All authors have read and agreed to the published version of the manuscript.

Funding: This research was funded by the National Key R&D Program of China, grant numbers 2019YFC1604801 and 2018YFC1602402.

Acknowledgments: Xianjiang Li and Rui Weng thank all the authors for their excellent contributions. The efforts of the reviewers contributed greatly to the quality of this Special Issue. We were very satisfied with the review process and management of the Special Issue, and we are grateful to all those involved.

Conflicts of Interest: The authors declare no conflict of interest.

References

1. Toxic Mineral Oil Found in Food Products. Available online: https://www.foodwatch.org/en/news/2021/toxic-mineral-oil-found-in-food-products/ (accessed on 18 October 2022).
2. Li, X.; Ma, W.; Li, H.; Zhang, Q.; Ma, Z. Determination of residual fipronil and its metabolites in food samples: A review. *Trends Food Sci. Tech.* **2020**, *97*, 185–195. [CrossRef]
3. Yang, B.; Ma, W.; Wang, S.; Shi, L.; Li, X.; Ma, Z.; Zhang, Q.; Li, H. Determination of eight neonicotinoid insecticides in Chinese cabbage using a modified QuEChERS method combined with ultra performance liquid chromatography-tandem mass spectrometry. *Food Chem.* **2022**, *387*, 132935. [CrossRef] [PubMed]
4. Pang, X.; Liu, X.; Peng, L.; Chen, Z.; Qiu, J.; Su, X.; Yu, C.; Zhang, J.; Weng, R. Wide-scope multi-residue analysis of pesticides in beef by ultra-high-performance liquid chromatography coupled with quadrupole time-of-flight mass spectrometry. *Food Chem.* **2021**, *351*, 129345. [CrossRef] [PubMed]
5. Weng, R.; Lou, S.; Pang, X.; Song, Y.; Su, X.; Xiao, Z.; Qiu, J. Multi-residue analysis of 126 pesticides in chicken muscle by ultra-high-performance liquid chromatography coupled to quadrupole time-of-flight mass spectrometry. *Food Chem.* **2020**, *309*, 125503. [CrossRef]
6. Weng, R.; Sun, L.; Jiang, L.; Li, N.; Ruan, G.; Li, J.; Du, F. Electrospun Graphene Oxide–Doped Nanofiber-Based Solid Phase Extraction Followed by High-Performance Liquid Chromatography for the Determination of Tetracycline Antibiotic Residues in Food Samples. *Food Anal. Methods* **2019**, *12*, 1594–1603. [CrossRef]
7. Li, X.; Ma, W.; Zhang, Q.; Li, H.; Liu, H. Determination of patulin in apple juice by amine-functionalized solid-phase extraction coupled with isotope dilution liquid chromatography tandem mass spectrometry. *J. Sci. Food Agric.* **2021**, *101*, 1767–1771. [CrossRef]
8. Ma, W.; Yang, B.; Li, J.; Li, X. Development of a Simple, Underivatized Method for Rapid Determination of Free Amino Acids in Honey Using Dilute-and-Shoot Strategy and Liquid Chromatography-Tandem Mass Spectrometry. *Molecules* **2022**, *27*, 1056. [CrossRef] [PubMed]
9. Zhao, Z.; Guo, Y.; Wei, J.; Chen, Q.; Chen, X. Fluorescent Copper Nanoclusters for Highly Sensitive Monitoring of Hypoxanthine in Fish. *J. Anal. Test.* **2021**, *5*, 76–83. [CrossRef]
10. Liu, P.; Weng, R.; Xu, Y.; Pan, Y.; Wang, B.; Qian, Y.; Qiu, J. Distinct Quality Changes of Garlic Bulb during Growth by Metabolomics Analysis. *J. Agric. Food. Chem.* **2020**, *68*, 5752–5762. [CrossRef] [PubMed]
11. Li, X.; Wang, S.; Guo, Z.; Li, X.; Zhang, Q.; Li, H. Determination of Fosetyl-Aluminum in Wheat Flour with Extract-Dilute-Shoot Procedure and Hydrophilic Interaction Liquid Chromatography Tandem Mass Spectrometry. *Separations* **2021**, *8*, 197. [CrossRef]
12. Tong, K.; Xie, Y.; Huang, S.; Liu, Y.; Wu, X.; Fan, C.; Chen, H.; Lu, M.; Wang, W. QuEChERS Method Combined with Gas- and Liquid-Chromatography High Resolution Mass Spectrometry to Screen and Confirm 237 Pesticides and Metabolites in Cottonseed Hull. *Separations* **2022**, *9*, 91. [CrossRef]
13. Wu, X.; Tong, K.; Yu, C.; Hou, S.; Xie, Y.; Fan, C.; Chen, H.; Lu, M.; Wang, W. Development of A High-Throughput Screening Analysis for 195 Pesticides in Raw Milk by Modified QuEChERS Sample Preparation and Liquid Chromatography Quadrupole Time-of-Flight Mass Spectrometry. *Separations* **2022**, *9*, 98. [CrossRef]
14. Bie, R.; Zhang, J.; Wang, Y.; Jin, D.; Yin, R.; Jiang, B.; Cao, J. Analysis of Multiclass Pesticide Residues in Tobacco by Gas Chromatography Quadrupole Time-of-Flight Mass Spectrometry Combined with Mini Solid-Phase Extraction. *Separations* **2022**, *9*, 104. [CrossRef]
15. Yang, Y.; He, Z.; Mu, L.; Xie, Y.; Wang, L. Simultaneous Determination of 23 Mycotoxins in Broiler Tissues by Solid Phase Extraction UHPLC-Q/Orbitrap High Resolution Mass Spectrometry. *Separations* **2021**, *8*, 236. [CrossRef]
16. Xiao, Y.; Liu, S.; Gao, Y.; Zhang, Y.; Zhang, Q.; Li, X. Determination of Antibiotic Residues in Aquaculture Products by Liquid Chromatography Tandem Mass Spectrometry: Recent Trends and Developments from 2010 to 2020. *Separations* **2022**, *9*, 35. [CrossRef]
17. Li, X.; Ma, Q.; Wei, C.; Cai, W.; Chen, H.; Xing, R.; Song, P. Green and Simple Extraction of Arsenic Species from Rice Flour Using a Novel Ultrasound-Assisted Enzymatic Hydrolysis Method. *Separations* **2022**, *9*, 105. [CrossRef]
18. Zhao, Y.; Wang, M.; Yang, M.; Zhou, J.; Wang, T. Determination of Selenomethionine, Selenocystine, and Methylselenocysteine in Egg Sample by High Performance Liquid Chromatography—Inductively Coupled Plasma Mass Spectrometry. *Separations* **2022**, *9*, 21. [CrossRef]
19. Li, X.; Wang, W.; Sun, S.; Wang, J.; Zhu, J.; Liang, F.; Zhang, Y.; Hu, G. Quantitative Analysis of Anthocyanins in Grapes by UPLC-Q-TOF MS Combined with QAMS. *Separations* **2022**, *9*, 140. [CrossRef]

Article

Determination of Fosetyl-Aluminum in Wheat Flour with Extract-Dilute-Shoot Procedure and Hydrophilic Interaction Liquid Chromatography Tandem Mass Spectrometry

Xianjiang Li *, Sheng Wang, Zhen Guo, Xiuqin Li, Qinghe Zhang and Hongmei Li

Food Safety Laboratory, Division of Metrology in Chemistry, National Institute of Metrology, Beijing 100029, China; wangsh@nim.ac.cn (S.W.); guozh@nim.ac.cn (Z.G.); lixq@nim.ac.cn (X.L.); zhangqh@nim.ac.cn (Q.Z.); lihm@nim.ac.cn (H.L.)
* Correspondence: lixianjiang@nim.ac.cn; Tel.: +86-10-64524737

Abstract: Fosetyl-aluminum is a widely used ionic fungicide. This pesticide is not amenable to the common multi-residue sample preparation methods. Herein, this paper describes a novel method for the simple and sensitive determination of fosetyl-aluminum residue in wheat flour. The sample preparation method involved extraction with water under ultrasonication and subsequent dilution with six-fold acetonitrile. The fosetyl-aluminum concentration was determined by hydrophilic interaction liquid chromatography tandem mass spectrometry. The limit of detection and quantification were only 5 and 10 ng/g, respectively, which meet the requirement of the current European legislation. Matrix-matched linearity ($r^2 = 0.9999$) was established in the range of 10–2000 ng/g. Satisfactory recoveries were achieved in the range of 95.6% to 105.2% for three levels of spiked samples (10, 50, and 100 ng/g). Finally, the method was applied to analyzing 75 wheat flour samples produced in four provinces in China. Two samples were positive with concentrations over the limit of detection. This is the first method focusing on fosetyl-aluminum determination in wheat flour with an extract-dilute-shoot strategy and is very promising for the routine quality control of fosetyl-aluminum in similar cereal matrices.

Keywords: extract-dilute-shoot; fosetyl-aluminum; hydrophilic interaction liquid chromatography; wheat flour

1. Introduction

Fosetyl-aluminum (fosetyl-Al) is a polar fungicide, and a replacement for the banned sodium arsenite [1]. This fungicide is widely used to control rot in plant roots. Due to its toxic effects, the Joint Meeting on Pesticide Residues recommended set the acceptable daily intake for fosetyl-Al to 0–1 mg/kg bw per day in 2017 [2]. The maximum residue level of fosetyl-Al in food is strictly controlled by Regulation 396/2005 of the European Commission [3], GB 2763-2021 of China [4], and 40CFR180.415 of the USA [5].

Fosetyl-Al is a highly polar compound with an ionic structure. This highly polar pesticide is not amenable to the common multi-residue sample preparation methods because it is difficult to partition into common organic solvents [6] and needs dedicated chromatographic conditions. In addition, it is difficult to retain fosetyl-Al in typical reverse-phase liquid chromatography and the co-eluted salts and polar matrix components seriously interfere with fosetyl-Al determination [7]. Therefore, there is an urgent demand for the development of a simple and general method that could be used to detect this "orphan pesticide". Until now, few solutions that address this problem have been reported, such as applying a specific column with a polar stationary phase or using an ion-pair reagent within the mobile phase [7]. Because it lacks UV absorption and fluorescence, fosetyl-Al is rather difficult to determine by conventional liquid chromatography detectors such as diode array detectors. In recent years, the use of liquid chromatography mass spectrometry (LC-MS) for analyte determination has aroused great attention due to its inherent selectivity and sensitivity [8]. Several recent works have described the use of LC-MS for fosetyl-Al determination in different food commodities, such as lettuce [7], tomato [9], grape [1], mango [10], olive oil [11], oat, and soy beans [12]. In these LC-MS methods, polar pesticides are separated with graphitized carbon (Hypercarb) columns [13] or hydrophilic interaction liquid chromatography (HILIC) columns [9,14,15]. To the best of our knowledge, no work has reported the determination of fosetyl-Al in wheat flour with the HILIC-MS/MS method.

Wheat flour is a staple food in many countries. It is a typical matrix that is high in carbohydrates and proteins and belongs in category 5 of the AOAC Food Triangle. During its long-term storage, fosetyl-Al may be illegally spiked into wheat flour to avoid fungus growth. Scarce literature is available for the analysis of this polar pesticide in wheat flour. Therefore, the State Administration for Market Regulation of China set up a project and supported us in developing a method for fosetyl-Al determination in wheat flour. Effective removal of the polar, soluble carbohydrates and proteins poses a great challenge in the sample preparation process. For the LC-MS method, sample preparation is necessary prior to analyte detection to eliminate interferences [16] and a lower matrix effect (ME) for MS determination [17–19], such as the hydrophilic-lipophilic balance cartridge [1] and the anion-exchange column [20]. One widely adopted sample preparation method was developed by the European Reference Laboratory to determine highly polar pesticides in plant-derived foods such as soy flour, named the Quick Polar Pesticides (QuPPe) method [21]. Although the method is capable of extracting various polar analytes, the extracts probably contain large amounts of matrix interferences that would contaminate the instruments. Furthermore, this QuPPe strategy is tedious and time-consuming. On the contrary, the extract-dilute-shoot procedure is a promising strategy for its simple and fast operation. As an improved method from dilute-and-shoot [22], this method is fit for solid samples. In detail, solid samples are extracted with a suitable extraction solvent to facilitate the migration of analytes into the liquid phase. Then, the liquid extract was diluted with a suitable dilution solvent before the shoot step. The dilution minimizes the ME and thus, reduces the need for an additional clean-up procedure. Recently, this procedure has gained increased attention in food analysis, including rice [23], tomato [24], fruit jam [25], and gingerbread [26]. However, the HILIC-MS/MS method with the extract-dilute-shoot procedure has not yet been applied for fosetyl-Al determination.

The aim of this work was to develop a simple and sensitive analytical method for fosetyl-Al determination in wheat flour. Therefore, we adopted an extract-dilute-shoot strategy for sample preparation, and combined it with HILIC separation and tandem mass spectrometry determination. To the best of our knowledge, this is the first report that focuses on fosetyl-Al determination in wheat flour with extract-dilute-shoot sample preparation and HILIC-MS/MS determination.

2. Materials and Methods

2.1. Chemicals and Materials

HPLC-grade acetonitrile and methanol were bought from Merck (Darmstadt, Germany). Water was supplied by Hangzhou Wahaha Group Co., Ltd. (Hangzhou, China). Ammonium formate and formic acid (FA) were provided by Sigma-Aldrich (St. Louis, MO, USA). Analyte-grade fosetyl-Al standard was purchased from Dr. Ehrenstorfer (Augsburg, Germany). All wheat flour samples were commercial products from the local market.

For the syringe filter (0.22 μm), GHP was from PALL Life Sciences (Ann Arbor, MI, USA), nylon was from UA Filter & Chrom (Taibei, Taiwan, China), PTFE PVDF and PP were from Jinteng Experimental Equipment (Tianjin, China). The syringe (2 mL) was from Jiangsu Yuzhi Medical instrument (Taixing, China).

Chromatography columns were Hypercarb columns (2.1 × 100 mm, 5 μm) from Thermo Fisher Scientific (Bellefonte, PA, USA); BEH C_{18} column (2.1 mm × 50 mm, 1.7 μm) and BEH amide column (2.1 × 50 mm, 1.7 μm) were obtained from Waters (Milford,

CT, USA). Guard columns of the same stationary phase were connected in front of the separation columns. All the columns were preconditioned according to the manufacturer's instructions before use.

2.2. Preparation of Calibration Solutions

A stock solution of fosetyl-Al was prepared by dissolving an accurately weighed portion of the pure standard compound in acetonitrile at a concentration of 250 µg/g. Blank wheat flour samples were spiked with the fosetyl-Al stock solution to prepare matrix-matched external calibration solutions. The concentrations of fosetyl-Al varied between 5 and 2000 ng/g. Then, the spiked samples were extracted by water, and the extracts were diluted with 6-fold acetonitrile. After centrifugation and filtration, the matrix-matched standards were ready for further use. All solutions were stored at $-20\ °C$ in the dark.

2.3. Instruments

Samples were centrifuged by a Hettich universal 320R centrifuge to separate the supernatant (Tuttlingen, Germany). Samples and solutions were mixed by a Vortex Genie-2 from Scientific Industries Inc. (New York, NY, USA). Samples were sonicated with a Branson 8510 Ultrasonic Cleaner (Danbury, CT, USA).

HILIC-MS/MS experiments were performed using a Waters instrument. An ACQUITY UPLC® system was used for LC separation. The LC system was connected to a triple quadrupole MS (TQ-S, Manchester, UK) with a Z-spray electrospray ionization interface. MassLynx™ 4.1 software (Milford, CT, USA) was used for instrument control and data acquisition. Nitrogen was used as the nebulizer gas and was supplied by the generator NM31LA of Peak Scientific (Scotland, UK).

2.4. Optimization of Extract-Dilute-Shoot Parameters

To accurately quantify fosetyl-Al concentration in wheat flour, it was important to extract fosetyl-Al and remove matrix interference by efficient sample preparation. To obtain optimal extraction efficiency for fosetyl-Al, several important parameters influencing the extraction efficiency were evaluated in this study, such as type and volume of extraction solvent, dilution factor of the extract, and type of filter. To obtain a reliable result, all the optimizations were carried out in fosetyl-Al-spiked wheat flour at a concentration of 50 ng/g in triplicate.

2.5. Sample Pretreatment

A 2.00 g portion of wheat flour was accurately weighed into a 50 mL polypropylene centrifuge tube, and mixed with 12 mL water using an automatic pipette. The tube was capped tightly and vortexed for 2 min to form a homogeneous paste. Then, the paste was extracted under sonication for 15 min at room temperature. Afterward, the mixture was vortexed again for 1 min. Then, the tube was centrifuged at 9000 rpm for 3 min, and 1 mL supernatant was collected and diluted with 6-fold acetonitrile. Finally, the extract was filtered with a 0.22 µm GHP syringe filter before analysis by HILIC-MS/MS.

2.6. Hydrophilic Interaction Liquid Chromatography Tandem Mass Spectrometry

Three different columns were tested for the retention of fosetyl-Al, including BEH C_{18}, BEH amide, and Hypercarb. Afterward, the mobile phase and additive were also optimized to achieve better separation and peak shape in liquid chromatography.

The ion source parameters were optimized automatically by a TQ-S system (Waters, Manchester, UK) with the direct infusion of fosetyl-Al solution at a flow rate of 10 µL/min, source temperature of 150 °C, desolvation temperature of 500 °C, capillary voltage of −2.3 kV, corn voltage of 30 V, source offset voltage of 50 V, desolvation gas at 700 L/h, cone gas at 150 L/h, and collision gas at 0.13 mL/min. The detection of fosetyl-Al was performed in multiple reactions monitoring mode with a collision energy of 10. The precursor ion with *m/z* 109 corresponded to the fosetyl anion. The product ion with *m/z* 81 was the product of the McLaffery rearrangement with the loss of ethene, and the product ion with m/z 63 was PO_2^- [7]. Therefore, the most intense transition of m/z 109 > m/z 81 was selected for quantification and the transition of m/z 109 > m/z 63 was used for qualification. Two transitions were selected to qualitatively and quantitatively detect fosetyl-Al in the validation study.

2.7. Method Validation

To validate the applicability of the developed method for fosetyl-Al determination, the linearity, linear range, limit of detection (LOD), limit of quantification (LOQ), ME, recovery, and precision were investigated. A calibration curve for the quantitative analysis was established using the matrix-matched standard of spiked wheat flour samples. The standards were produced using the developed extract-dilute-shoot method. The spike concentrations were in the range of 5–2000 ng/g. The linearity of the method was evaluated by using a linear regression curve fit with the areas obtained for the matrix-matched standard. The LOD and LOQ were calculated from chromatograms of fortified samples according to SANTE/2020/12830, defined as the concentration of signal-to-noise ratios larger than 3 and 10, respectively. To quantitively evaluate the ME (suppression or enhancement), standard solutions of both solvent and matrix were shot into the HILIC-MS/MS. The ME was assessed by the slope of the calibration curve between the matrix and solvent standards by the below equation [27]. An ME of less than 100% indicates matrix suppression; an ME greater than 100% indicates matrix enhancement. To establish the reliability and validity of the analytical method, recovery and precision tests of fosetyl-Al were carried out in blank samples fortified with three different levels (10, 50, and 100 ng/g) with five replicates.

$$ME = b_m/b_s \times 100\%$$

where b_m and b_s are the angular coefficients of the curve in the matrix and in the solvent, respectively.

2.8. Real Sample Analysis

Seventy-five commercial wheat flour samples were collected from four provinces of China, including Shandong (25 samples), Henan (15 samples), Hebei (3 samples), and Jiangsu (32 samples). Two replicates were tested from each sample to ensure that reliable results were collected. Retention time and intensity of product ions were used to identify positive samples. Quantification was achieved using an external matrix-matched calibration curve that was produced from the peak area of fosetyl-Al versus the corresponding concentration of the spiked wheat flour samples.

3. Results and Discussion

3.1. Extract-Dilute-Shoot Optimization

3.1.1. Type of Extraction Solvent

The choice of extraction solvent directly impacted the extraction efficiency of fosetyl-Al for further analysis. Initially, acidified methanol, described by the QuPPe method, was used for the extraction of the fosetyl-Al. However, this extraction method was not suitable to the wheat flour matrix, because a doughy mixture was inevitably generated when the acidified methanol was added. The doughy substance was difficult to remove from the solution via high-speed centrifugation or syringe filter. This finding is consistent with a previous study about the analysis of fosetyl-Al in soy nutraceuticals [28]. To achieve a satisfactory extraction efficiency, five kinds of polar extraction solvents were tested in this study. The extraction was evaluated by recoveries of spiked fosetyl-Al. From the results in Figure 1A, polarity played a dominated role in the extraction; therefore, recoveries were in the following order: water > water/methanol (50%) > methanol > water (0.5% FA) > isopropanol > acetonitrile. This indicated that water had excellent performance for the extraction of fosetyl-Al. In water, fosetyl-Al transforms into fosetyl anions. The addition of FA would turn the negative ion to neutral molecular and lower extraction efficiency. Therefore, water was used in the following extractions, consistent with a previous report [29].

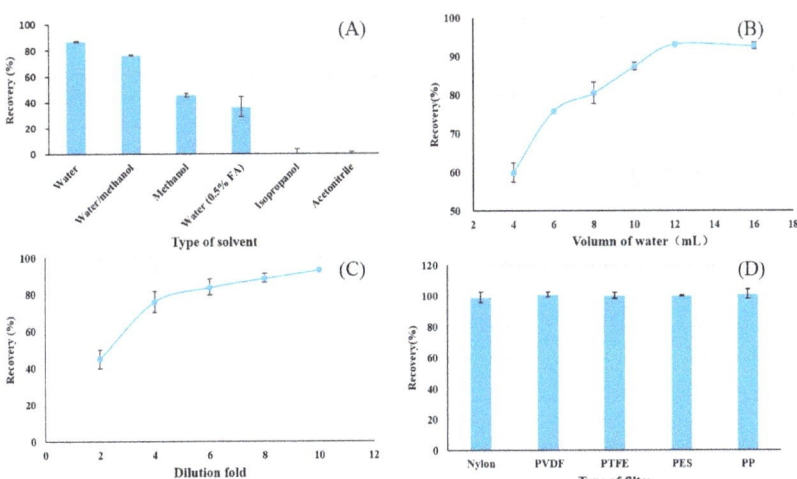

Figure 1. Optimization of the (**A**) type and (**B**) volume of extraction solvent; (**C**) effect of dilution fold; (**D**) type of filter (n = 3).

3.1.2. Volume of Extraction Solvent

The volume of water can directly influence the extraction efficiency of wheat flour. A larger volume of extraction solvent would extract more fosetyl-Al in theory, while signal intensity may decrease remarkably because of the dilution effect. On the other hand, a smaller volume would not provide enough quantity of the sample for LC-MS analysis and the reproducibility would be poor. Herein, the volume of water was optimized from 4 L to 16 mL for 2.00 g wheat flour to find the optimal volume. From Figure 1B, the recovery increased with water volume from 4 to 12 mL, and no significant signal enhancement was observed with increasing water volume. Better recovery would provide better accuracy and that was very important for the real samples test. Moreover, larger volume led to less variation. As a result, 12 mL was chosen for following experiment.

3.1.3. Effect of Dilution Fold

In wheat flour, the main compositions were 71.2% carbohydrate, 15.1% protein, 2.7% lipid, and 9.4% water [30]. During the water extraction, some soluble carbohydrates and proteins are extracted simultaneously. Therefore, acetonitrile was chosen to dilute the extract solution to precipitate these compounds and minimize the ME. Acetonitrile is widely used for protein precipitation [31] and it can lower the solubility of carbohydrates [32]. Once acetonitrile was added, the extract turned cloudy immediately. The employed dilution factors usually range from 2 to 50 according to the matrix [22]. Thus, dilution fold was optimized from 2 to 10, as exhibited in Figure 1C. Recoveries increased with higher dilution and relative standard deviations (RSDs) decreased. This indicated that lower ME and better repeatability were achieved with acetonitrile dilution. Conversely, large dilution factors would subsequently reduce the sensitivity of the method in terms of LOD. In conclusion, a six-fold dilution factor was selected as a compromise between sensitivity and repeatability.

3.1.4. Type of Filter

To remove the cloudy particles in solution and prolong the lifetime of the column, filtration is necessary before LC separation. Sometimes, pesticides can be adsorbed by the membrane filters and lead to analyte loss and low recovery [33]. As a result, filter adsorption of fosetyl-Al was evaluated among five widely used filters (GHP, nylon, PTFE, PVDF, and PP). From the results in Figure 1D, there was no significant difference between the tested filters. So, any kind of the tested filter was feasible for further tests.

3.2. Hydrophilic Interaction Liquid Chromatography Tandem Mass Spectrometry

The column was selected on the basis of the polarity of fosetyl-Al. As shown in Figure 2, preliminary testing with the BEH C_{18} column demonstrated its low retention capability with fosetyl-Al, which appeared in the peak void with a retention time of 0.35 min. Due to the highly polar nature of fosetyl-Al, it was difficult to obtain enough retention on a C_{18} column. Conversely, the Hypercarb column exhibited such a strong retention for fosetyl-Al that it resulted in an increased elution time. The BEH amide column had better retention behavior toward fosetyl-Al, with 17 times higher sensitivity than the Hypercarb column. With regard to the optimization of the mobile phase, acetonitrile/water showed better elution performance than methanol/water on the BEH amide column. Moreover, to keep the pH value and the retention time of fosetyl anion constant throughout the run, the buffer concentration was further optimized with gradient elution. Referring to previous research on polar pesticides [34], mobile phase A was chosen as water with 5.0 mmol/L ammonium formate and B was acetonitrile. The gradient started at 90% of phase B to elute non-polar compounds while fosetyl-Al was still retained. Afterward, phase B decreased linearly to 50% over 3 min and the polar compounds eluted gradually. Then, the mobile phase composition returned to the initial condition in 0.1 min and was held for 1.9 min for re-equilibration. The separation was operated at a flow rate of 0.3 mL/min and the column temperature was kept at 35 °C. The total chromatographic run time was 5 min. The retention time of fosetyl-Al was 1.86 min under this condition.

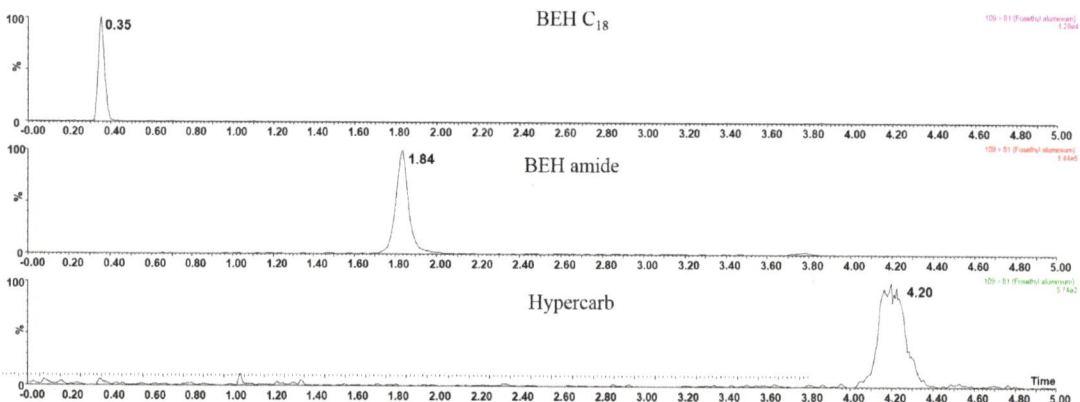

Figure 2. Performance comparison between three types of columns.

3.3. Method Validation

As listed in Table 1, the linear range of the developed method covered from 10 to 2000 ng/g with eight concentration levels in the matrix-matched standards. Simultaneously, correlation coefficient (r^2) of the matrix-matched calibration curve equaled 0.9999, which is very satisfactory for accurate quantification. The LOD and LOQ of the developed method were 5 and 10 ng/g, respectively, using the signal-to-noise ratio of the qualifier transition signal. This is sufficiently low to meet the maximum residue limits for many regulations, including the European Commission, China, and the USA. The proposed method showed satisfactory accuracy with recoveries of 95.6%, 105.2%, and 104.6%, respectively. Method precision was evaluated by the RSDs of five repetitions. The result was lower than 6.2%, indicating that this method is quite fit for routine analysis. The ME was calculated by comparing the slopes of standards prepared in wheat flour extract and water/acetonitrile solvent. The ME evaluation (88.33%) showed that the residual matrix could suppress the fosetyl-Al signal. Therefore, a matrix-matched calibration solution was used for accurate quantitation during the real sample analysis.

Table 1. Validation data of the developed method for the detection of fosetyl-Al.

Linear Range (ng/g)	Linearity (r^2)	LOD (ng/g)	ME	RSDs% (n = 5)	Recovery		
					10 ng/g	50 ng/g	100 ng/g
10–2000	0.9999	5	88.33%	6.2	95.6%	105.2%	104.6%

From the above, a simple and sensitive extract-dilute-shoot HILIC-MS/MS method was established. This proposed method is easy to operate and has excellent sensitivity (Table 2), which is promising for the analysis of fosetyl-Al in wheat flour samples.

Table 2. Method comparison of fosetyl-Al in different matrices.

Matrix	Sample Preparation	ME (%)	LOD (ng/g)	LOQ (ng/g)	Linear Range (ng/g)	Linearity	Ref.
Grape	SPE	13	9	29	10–1000 [1]	0.995	[1]
Lettuce	Water extraction, 5-fold dilution	/	50	200	5–10,001	0.9993	[7]
Tomato	QuPPe method	137	/	25	25–1000	0.999	[9]
Mango	QuPPe method	70	/	50	50–2000	0.99	[10]
Olive oil	QuPPe method	80	/	10	10–1000	0.9932	[11]
Soy beans	Methanol extraction, 12.5-fold dilution	96	/	20	10–10,000	0.9969	[12]
Wheat flour	Water extraction, 6-fold dilution	88.33	5	10	10–2000	0.9999	This work

[1] ng/mL.

3.4. Real Sample Analysis

In order to study the applicability of the proposed method, the developed method was applied to analyze the fosetyl-Al residual in commercial wheat flour samples. In total, 75 samples were collected from four provinces that covered China's main wheat flour production areas. For each kind of sample, the determination was repeated two times. Matrix-matched calibrations were injected in every sequence of samples in order to explore the carry-over effect and to ensure that a reagent blank was injected immediately after the highest standard. As a result, two samples were positive for fosetyl-Al, as shown in Figure 3. The signal intensity reflected that their concentrations were higher than the LOD and lower than the LOQ. One sample was produced in Shandong Province, the other was from Henan Province. All samples were below the maximum residue limits, suggesting that it is generally safe to consume wheat flour in China.

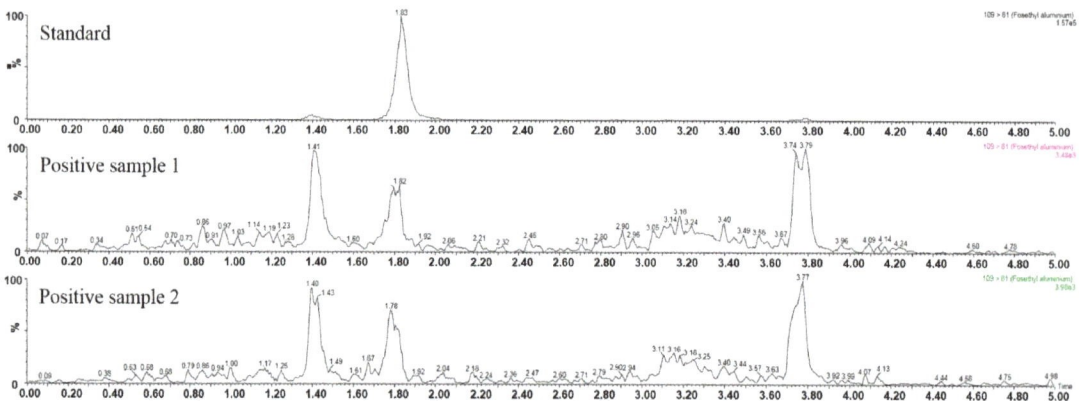

Figure 3. Chromatograms of the positive samples.

4. Conclusions

In this study, we developed and validated a method by combining an extract-dilute-shoot strategy for sample preparation with HILIC-MS/MS for fosetyl-Al determination in wheat flour. This simple and fast dilution procedure effectively lowered the ME and improved the repeatability. Moreover, HILIC column provided both sufficient retention of fosetyl-Al and removal of the matrix, and MS/MS measurement demonstrated excellent sensitivity and selectivity. This approach offers simple operation, minimal consumption of chemicals, wide linear range, and high sensitivity. Satisfactory results were obtained as evidenced by matrix-matched standards. The observed LOQ was 10 ng/g in wheat flour. The satisfactory precisions and recoveries achieved with the spiked samples demonstrated the reliability and practicability of the developed method. The proposed method was applied to the quantitation of fosetyl-Al in 75 wheat flour samples, which overall showed low levels of contamination. These results showed that our proposed method is very promising for routine analysis and could find more applications in quality control of fosetyl-Al in high-carbohydrate food matrices.

Author Contributions: Conceptualization, methodology, validation, and writing—original draft preparation, X.L. (Xianjiang Li); data curation and writing—review and editing, S.W.; investigation, Z.G.; supervision, X.L. (Xiuqin Li); project administration, Q.Z.; funding acquisition, H.L. All authors have read and agreed to the published version of the manuscript.

Funding: This research was funded by the National Key R&D Program of China, grant number 2019YFC1604801.

Institutional Review Board Statement: Not applicable.

Informed Consent Statement: Not applicable.

Conflicts of Interest: The authors declare no conflict of interest.

References

1. Chamkasem, N. Determination of Glyphosate, Maleic Hydrazide, Fosetyl Aluminum, and Ethephon in Grapes by Liquid Chromatography/Tandem Mass Spectrometry. *J. Agric. Food. Chem.* **2017**, *65*, 7535–7541. [CrossRef] [PubMed]
2. *Pesticide Residues in Food*; Joint FAO/WHO Meeting on Pesticide Residues: Geneva, Switzerland, 2017. Available online: http://www.fao.org/fileadmin/templates/agphome/documents/Pests_Pesticides/JMPR/Report2017/web_2017_JMPR_Report_Final.pdf (accessed on 20 September 2021).
3. European Food Safety Authority. Modification of the existing maximum residue level for fosetyl in blackberry, celeriac and Florence fennel. *EFSA J.* **2015**, *13*, 4327. [CrossRef]
4. NHC; MoA; SAMR of China. *National Food Safety Standard—Maximum Residue Limits for Pesticides in Food*; China Agriculture Press: Beijing, China, 2020.

5. Aluminum Tris (O-ethylphosphonate); Tolerances for Residues. Available online: https://www.ecfr.gov/cgi-bin/text-idx?SID=c30d88a8176fe20021d927e53e9422c7&mc=true&node=se40.26.180_1415&rgn=div8 (accessed on 20 September 2021).
6. Buiarelli, F.; Di Filippo, P.; Riccardi, C.; Pomata, D.; Marsiglia, R.; Console, C.; Puri, D. Hydrophilic Interaction Liquid Chromatography-Tandem Mass Spectrometry Analysis of Fosetyl-Aluminum in Airborne Particulate Matter. *J. Anal. Methods Chem.* **2018**, *2018*, 8792085. [CrossRef] [PubMed]
7. Hernández, F.; Sancho, J.V.; Pozo, Ó.J.; Villaplana, C.; Ibáñez, M.; Grimalt, S. Rapid Determination of Fosetyl-Aluminum Residues in Lettuce by Liquid Chromatography/Electrospray Tandem Mass Spectrometry. *J. AOAC Int.* **2003**, *86*, 832–838. [CrossRef] [PubMed]
8. Xu, Y.; Wang, H.; Li, X.; Zeng, X.; Du, Z.; Cao, J.; Jiang, W. Metal–organic framework for the extraction and detection of pesticides from food commodities. *Compr. Rev. Food Sci. F.* **2021**, *20*, 1009–1035. [CrossRef]
9. Manzano-Sanchez, L.; Martinez-Martinez, J.A.; Dominguez, I.; Martinez Vidal, J.L.; Frenich, A.G.; Romero-Gonzalez, R. Development and Application of a Novel Pluri-Residue Method to Determine Polar Pesticides in Fruits and Vegetables through Liquid Chromatography High Resolution Mass Spectrometry. *Foods* **2020**, *9*, 553. [CrossRef]
10. Marinho Pereira da Silva, H.C.; Galindo Bedor, D.C.; Cunha, A.N.; dos Santos Rodrigues, H.O.; Telles, D.L.; Pessoa Araujo, A.C.; de Santana, D.P. Ethephon and fosetyl residues in fruits from São Francisco Valley, Brazil. *Food Addit. Contam. B* **2019**, *13*, 16–24. [CrossRef]
11. Nortes-Mendez, R.; Robles-Molina, J.; Lopez-Blanco, R.; Vass, A.; Molina-Diaz, A.; Garcia-Reyes, J.F. Determination of polar pesticides in olive oil and olives by hydrophilic interaction liquid chromatography coupled to tandem mass spectrometry and high resolution mass spectrometry. *Talanta* **2016**, *158*, 222–228. [CrossRef]
12. López, S.H.; Scholten, J.; Kiedrowska, B.; de Kok, A. Method validation and application of a selective multiresidue analysis of highly polar pesticides in food matrices using hydrophilic interaction liquid chromatography and mass spectrometry. *J. Chromatogr. A* **2019**, *1594*, 93–104. [CrossRef]
13. Savini, S.; Bandini, M.; Sannino, A. An Improved, Rapid, and Sensitive Ultra-High-Performance Liquid Chromatography-High-Resolution Orbitrap Mass Spectrometry Analysis for the Determination of Highly Polar Pesticides and Contaminants in Processed Fruits and Vegetables. *J. Agric. Food. Chem.* **2019**, *67*, 2716–2722. [CrossRef] [PubMed]
14. Dias, J.; López, S.H.; Mol, H.; de Kok, A. Influence of different hydrophilic interaction liquid chromatography stationary phases on method performance for the determination of highly polar anionic pesticides in complex feed matrices. *J. Sep. Sci.* **2021**, *44*, 2165–2176. [CrossRef] [PubMed]
15. Zhang, Q.; Yang, F.-Q.; Ge, L.; Hu, Y.-J.; Xia, Z.-N. Recent applications of hydrophilic interaction liquid chromatography in pharmaceutical analysis. *J. Sep. Sci.* **2017**, *40*, 49–80. [CrossRef] [PubMed]
16. Rezaei, S.M.; Makarem, S.; Alexovič, M.; Tabani, H. Simultaneous separation and quantification of acidic and basic dye specimens via a dual gel electro-membrane extraction from real environmental samples. *J. Iran. Chem. Soc.* **2021**, *18*, 2091–2099. [CrossRef]
17. Li, X.; Li, H.; Ma, W.; Guo, Z.; Li, X.; Li, X.; Zhang, Q. Determination of patulin in apple juice by single-drop liquid-liquid-liquid microextraction coupled with liquid chromatography-mass spectrometry. *Food Chem.* **2018**, *257*, 1–6. [CrossRef] [PubMed]
18. Li, X.; Ma, W.; Li, H.; Zhang, Q.; Ma, Z. Determination of residual fipronil and its metabolites in food samples: A review. *Trends Food Sci. Technol.* **2020**, *97*, 185–195. [CrossRef]
19. Gao, S.; Wu, G.; Li, X.; Chen, J.; Wu, Y.; Wang, J.; Zhang, Z. Determination of Triazine Herbicides in Environmental Water Samples by Acetonitrile Inorganic Salt Aqueous Two-Phase Microextraction System. *J. Anal. Test.* **2018**, *2*, 322–331. [CrossRef]
20. Rajski, L.; Diaz Galiano, F.J.; Cutillas, V.; Fernandez-Alba, A.R. Coupling Ion Chromatography to Q-Orbitrap for the Fast and Robust Analysis of Anionic Pesticides in Fruits and Vegetables. *J. AOAC Int.* **2018**, *101*, 352–359. [CrossRef] [PubMed]
21. Anastassiades, M.; Kolberg, D.I.; Wachtler, E.E.A.-K.; Benkenstein, A.; Zechmann, S.; Mack, D.; Wildgrube, C.; Barth, A.; Sigalov, I.; Görlich, S.; et al. *Quick Method for the Analysis of Numerous Highly Polar Pesticides in Food Involving Extraction with Acidified Methanol and LC-MS/MS Measurement I. Food of Plant Origin (QuPPe-PO-Method)*, 11th ed.; EU Reference Laboratory for Pesticides: Stuttgart, Germany, 2020. Available online: https://www.eurl-pesticides.eu/userfiles/file/EurlSRM/meth_QuPPe_PO_V11(1).pdf (accessed on 20 September 2021).
22. Greer, B.; Chevallier, O.; Quinn, B.; Botana, L.M.; Elliott, C.T. Redefining dilute and shoot: The evolution of the technique and its application in the analysis of foods and biological matrices by liquid chromatography mass spectrometry. *Trends Anal. Chem.* **2021**, *141*, 116284. [CrossRef]
23. da Silva, L.P.; Madureira, F.; de Azevedo Vargas, E.; Faria, A.F.; Augusti, R. Development and validation of a multianalyte method for quantification of mycotoxins and pesticides in rice using a simple dilute and shoot procedure and UHPLC-MS/MS. *Food Chem.* **2019**, *270*, 420–427. [CrossRef] [PubMed]
24. Gebrehiwot, W.H.; Erkmen, C.; Uslu, B. A novel HPLC-DAD method with dilute-and-shoot sample preparation technique for the determination of buprofezin, dinobuton and chlorothalonil in food, environmental and biological samples. *Int. J. Environ. Anal. Chem.* **2020**, *4*, 1–15. [CrossRef]
25. Petrarca, M.H.; Meinhart, A.D.; Godoy, H.T. Dilute-and-Shoot Liquid Chromatography Approach for Simple and High-throughput Analysis of 5-Hydroxymethylfurfural in Fruit-based Baby Foods. *Food Anal. Methods* **2020**, *13*, 942–951. [CrossRef]
26. Tölgyesi, Á.; Sharma, V.K. Determination of acrylamide in gingerbread and other food samples by HILIC-MS/MS: A dilute-and-shoot method. *J. Chromatogr. B* **2020**, *1136*, 121933. [CrossRef] [PubMed]

27. Mao, X.; Xiao, W.; Wan, Y.; Li, Z.; Luo, D.; Yang, H. Dispersive solid-phase extraction using microporous metal-organic framework UiO-66: Improving the matrix compounds removal for assaying pesticide residues in organic and conventional vegetables. *Food Chem.* **2021**, *345*, 128807. [CrossRef]
28. Alves, R.D.; Romero-González, R.; López-Ruiz, R.; Jiménez-Medina, M.L.; Frenich, A.G. Fast determination of four polar contaminants in soy nutraceutical products by liquid chromatography coupled to tandem mass spectrometry. *Anal. Bioanal. Chem.* **2016**, *408*, 8089–8098. [CrossRef] [PubMed]
29. Mol, H.G.J.; van Dam, R.C.J. Rapid detection of pesticides not amenable to multi-residue methods by flow injection–tandem mass spectrometry. *Anal. Bioanal. Chem.* **2014**, *406*, 6817–6825. [CrossRef]
30. Flour, Whole Wheat, Unenriched. Available online: https://fdc.nal.usda.gov/fdc-app.html#/food-details/790085/nutrients (accessed on 10 September 2021).
31. Li, X.; Li, H.; Ma, W.; Guo, Z.; Li, X.; Song, S.; Tang, H.; Li, X.; Zhang, Q. Development of precise GC-EI-MS method to determine the residual fipronil and its metabolites in chicken egg. *Food Chem.* **2019**, *281*, 85–90. [CrossRef]
32. Kozlik, P.; Molnarova, K.; Jecmen, T.; Krizek, T.; Goldman, R. Glycan-specific precipitation of glycopeptides in high organic content sample solvents used in HILIC. *J. Chromatogr. B* **2020**, *1150*, 122196. [CrossRef]
33. Doulia, D.S.; Anagnos, E.K.; Liapis, K.S.; Klimentzos, D.A. Removal of pesticides from white and red wines by microfiltration. *J. Hazard. Mater.* **2016**, *317*, 135–146. [CrossRef]
34. Chamkasem, N.; Harmon, T. Direct determination of glyphosate, glufosinate, and AMPA in soybean and corn by liquid chromatography/tandem mass spectrometry. *Anal. Bioanal. Chem.* **2016**, *408*, 4995–5004. [CrossRef]

Article

QuEChERS Method Combined with Gas- and Liquid-Chromatography High Resolution Mass Spectrometry to Screen and Confirm 237 Pesticides and Metabolites in Cottonseed Hull

Kaixuan Tong [1], Yujie Xie [1], Siqi Huang [2], Yongcheng Liu [2], Xingqiang Wu [1], Chunlin Fan [1], Hui Chen [1,*], Meiling Lu [3] and Wenwen Wang [3]

[1] Key Laboratory of Food Quality and Safety for State Market Regulation, Chinese Academy of Inspection and Quarantine, Beijing 100176, China; tongkaixuan9097@163.com (K.T.); happycch@163.com (Y.X.); xingqiangheda@163.com (X.W.); caiqfcl@163.com (C.F.)
[2] Laboratory of Heilongjiang Feihe Dairy Co., Ltd., Qiqihar 164800, China; mx19981001@163.com (S.H.); chuxia0xue@126.com (Y.L.)
[3] Agilent Technologies (China) Limited, Beijing 100102, China; mei-ling.lu@agilent.com (M.L.); wen-wen_wang@agilent.com (W.W.)
* Correspondence: chenh@caiq.org.cn

Citation: Tong, K.; Xie, Y.; Huang, S.; Liu, Y.; Wu, X.; Fan, C.; Chen, H.; Lu, M.; Wang, W. QuEChERS Method Combined with Gas- and Liquid-Chromatography High Resolution Mass Spectrometry to Screen and Confirm 237 Pesticides and Metabolites in Cottonseed Hull. *Separations* **2022**, *9*, 91. https://doi.org/10.3390/separations9040091

Academic Editor: Chiara Emilia Cordero

Received: 8 March 2022
Accepted: 31 March 2022
Published: 2 April 2022

Publisher's Note: MDPI stays neutral with regard to jurisdictional claims in published maps and institutional affiliations.

Copyright: © 2022 by the authors. Licensee MDPI, Basel, Switzerland. This article is an open access article distributed under the terms and conditions of the Creative Commons Attribution (CC BY) license (https://creativecommons.org/licenses/by/4.0/).

Abstract: Cottonseed hull is a livestock feed with large daily consumption. If pesticide residues exceed the standard, it is easy for them to be introduced into the human body through the food chain, with potential harm to consumer health. A method for multi-residue analysis of 237 pesticides and their metabolites in cottonseed hull was developed by gas-chromatography and liquid-chromatography time-of-flight mass spectrometry (GC-QTOF/MS and LC-QTOF/MS). After being hydrated, a sample was extracted with 1% acetic acid in acetonitrile, then purified in a clean-up tube containing 400 mg MgSO$_4$, 100 mg PSA, and 100 mg C18. The results showed that this method has a significant effect in removing co-extracts from the oily matrix. The screening detection limit (SDL) was in the range of 0.2–20 μg/kg, and the limit of quantification (LOQ) was in the range of 0.2–20 μg/kg. The recovery was verified at the spiked levels of 1-, 2-, and 10-times LOQ (n = 6), and the 237 pesticides were successfully verified. The percentages of pesticides with recovery in the range of 70–120% were 91.6%, 92.8%, and 94.5%, respectively, and the relative standard deviations (RSDs) of all pesticides were less than 20%. This method was successfully applied to the detection of real samples. Finally, this study effectively reduced the matrix effect of cottonseed hull, which provided necessary data support for the analysis of pesticide residues in oil crops.

Keywords: QuEChERS; gas-chromatography high resolution mass spectrometry; liquid-chromatography high resolution mass spectrometry; pesticide residues; cottonseed hull

1. Introduction

The composition of cottonseed hull is similar to that of soybean concentrate, with a high content of cellulose that can enhance the digestive systems of ruminants. Cottonseed hull has been widely used as an alternative feed for ruminants, due to its low price, easy availability, and excellent mixing performance [1–3]. The excessive and illegal use of pesticides during forage planting makes it easy for pesticides to enter the food chain and accumulate in animal adipose tissue [4], and human consumers may indirectly experience food safety problems through contact with livestock products. The composition of the oily matrix is complex: in addition to fat, it contains polysaccharides, proteins, pigments, and other substances. In the process of residue analysis, problems such as matrix enhancement, matrix inhibition, and retention-time shifts may occur in the detection of pesticides, which will hinder the detection of target compounds [5,6]. Therefore, it is urgent to develop a detection technique for the oily matrix to solve these problems.

The analysis of pesticide residue usually includes the following steps: (1) extraction of the target compound; (2) removal of interference from the extract; and (3) qualitative and quantitative detection of the target compound [4]. Lipophilic pesticides tend to be concentrated in fat. Improper pretreatment will affect the detection sensitivity, recovery, and sample throughput [7]. The current pretreatment methods for plant-derived oil substrates mainly include dispersion liquid-liquid micro-extraction (DLLME) [8], matrix solid phase dispersion (MSPD) [9,10], low temperature fat precipitation (LTFP) [11], solid phase extraction (SPE) [5], and QuEChERS [12–16]. The QuEChERS method requires fewer reagent consumables and short pretreatment time, so it is accepted by more and more experimenters [17]. Theurillat et al. established the QuEChERS method to determine the residues of various pesticides and verified the method for 176 pesticides in six oily matrices [12]. Rutkowska et al. investigated the matrix effect and recovery of four seed samples of cress, fennel, flax, and hemp. The final method verified 248 pesticides, and the LOQs reached 0.005 mg/kg [14]. Banerjee et al. used the QuEChERS method to analyze more than 220 pesticide residues in sesame seeds. This method can effectively reduce the interference of the matrix effect by freezing and degreasing at $-80\ °C$ and then purifying the oil.

The current trend of separation science is to develop new chromatographic mass spectrometry methods that can detect multiple compounds at the same time after a single injection, thereby reducing analysis time and cost [18]. The current detection technology for the detection of pesticide residues in oily matrices is mainly triple quadrupole mass spectrometry (MS/MS) [13,19–21]. The data was collected according to the specific nucleo-cytoplasmic ratio of the specified compound, but other compounds that were not in the list could not be identified. When analyzing a large number of compounds, the sensitivity and selectivity are limited. Due to their high resolution, precise mass accuracy, outstanding full-scan sensitivity, and complete mass spectrometry information, high-resolution mass spectrometry (HRMS), such as time-of-flight mass spectrometry (TOF/MS) and quadrupole Orbitrap mass spectrometry (Obitrap/MS), can be used without additional sample injection. Under retrospective analysis, with these advantages, HRMS has been widely used in the field of food analysis [22,23]. Lehotay et al. used GC-TOF to analyze 34 pesticides in flaxseed, dough, and peanuts [15]. Amadeo et al. used GC-QTOF to verify 166 pesticide residues in avocados and almonds [24].

To ensure the safety of livestock feed and to prevent pesticide residues from being introduced into the human body through the food chain, this work established a QuEChERS multi-residue analysis method, and used GC- and LC-QTOF/MS techniques to verify 237 pesticides in cottonseed hull. By optimizing the hydration volume, extraction solvent, salting-out agent, and clean-up sorbents, the influence of the matrix effect was reduced and the pesticide recovery was optimized. Finally, this method was successfully applied to the analysis of actual samples, providing data support for the risk of pesticide residues in oily substrate monitoring.

2. Materials and Methods

2.1. Chemicals and Reagents

Pesticide standards (purity $\geq 98\%$) were obtained from Tianjin Alta Scientific (Tianjin, China). Sodium chloride, magnesium sulfate, and sodium sulfate (analytical purity) were obtained from Tianjin Fuchen Chemical Reagent Ltd. (Tianjin, China). Primary secondary amine (PSA) and C18 were purchased from Agilent Technologies (Santa Clara, CA, USA). Methanol, acetonitrile, and toluene (chromatographic purity) were obtained from Anpel Laboratory Technology (Shanghai, China). Formic acid and ammonium acetate (mass spectrometry grade) were obtained from Honeywell (Muskegon, MI, USA).

2.2. Apparatus

HPLC-QTOF/MS Agilent 1290 and Agilent 6550 equipped with Agilent Dual Jet Stream ESI and GC-QTOF/MS Agilent 7890B and Agilent 7200 were obtained from Agilent Technologies (Santa Clara, CA, USA). A Milli-QTM Ultrapure Water System was obtained from Millipore (Milford, MA, USA). An N-EVAP112 Nitrogen Blowing Concentrator was obtained from Organomation Associates (Worcester, MA, USA). An AH-30 Automatic homogenizer was obtained from RayKol Group Corp., Ltd. (Xiamen, China). An MS204S Electronic Analytical Balance was obtained from Mettler Toledo (Shanghai, China).

2.3. Standard Solution

Ten mg of the standard substance was accurately weighed into a 10 mL brown volumetric flask. a suitable reagent was selected according to the solubility of the compound in the organic reagent. It was dissolved by ultrasound and diluted to the mark to a standard solution of 1 mg/L. The standard solution was stored at $-18\ °C$ in the dark. As needed, a pipette with an appropriate amount of the standard stock solution was diluted with methanol to prepare a working solution of appropriate concentration, and stored at 4 °C in the dark.

2.4. Sample Preparation Method

Based on other oily matrix sample preparation methods [12,16], a modified QuEChERS method was used for the detection of cottonseed hull. Two g (accurate to ± 0.01 g) of sample were transferred into a 50 mL centrifuge tube; 2 mL of ultrapure water were added for hydration and then extracted with 10 mL of 1% acetic acid in acetonitrile. The homogenizer was used to homogenize the sample for 1 min at $13,500\times g$; then, 4 g $MgSO_4$, 1 g NaCl and a ceramic homoproton were added. The mixture was shaken for 10 min and centrifuged at $3155\times g$ for 5 min; then, 3 mL of supernatant was transferred to a clean-up tube containing 400 mg $MgSO_4$, 100 mg PSA, and 100 mg C18. After shaking for 10 min and being centrifuged at $3155\times g$ for 5 min, 1 mL of supernatant was dried under nitrogen, then ultrasonically redissolved with ethyl acetate containing internal heptachlor-exo-epoxide for GC-QTOF/MS analysis, and ultrasonically redissolved with acetonitrile aqueous solution (2:3, v/v) containing internal standard atrazine D5 for LC-QTOF/MS analysis.

2.5. Instrument Parameters

The instrument parameters of LC-QTOF/MS and GC-QTOF/MS were configured according to a previous paper published by our laboratory [25].

An LC-QTOF/MS: ZORBAX SB-C18 column (100 mm × 2.1 mm, 3.5 µm, Agilent Technologies) was used for separation at 40 °C; 5 mmol/L ammonium acetate with 0.1% (v/v) formic acid aqueous solution and acetonitrile were applied as phase A and phase B. The flow rate was set at 0.4 mL/min. The gradient program was set as follows: 0 min, 1% B; 3 min, 30% B; 6 min, 40% B; 9 min, 40% B; 15 min, 60% B; 19 min, 90% B; 23 min, 90% B; 23.01 min, 1% B. The equilibrium time was 4 min. The injection volume was 5 µL.

The Agilent Dual Jet Stream (AJS) ESI source (Agilent Technologies) was set in positive full scan (m/z 50–1000) mode; the capillary voltage was 4 kV; nitrogen was used as the nebulizer gas at 0.14 MPa; the sheath gas temperature was set at 375 °C with 11.0 L/min; the drying gas flow rate was 12.0 L/min; the drying gas temperature was 225 °C; the fragmentation voltage was 345 V. In all ions Mass/Mass mode, the collision energy was 0 V at 0 min, and 0, 15, and 35 V at 0.5 min, respectively. The total program duration was 27.01 min.

GC-QTOF/MS: HP-5 MS UI (30 m × 0.25 mm, 0.25 µm, Agilent Technologies) was used for separation at 40 °C. The oven temperature gradient was started at 40 °C for 1 min, increased at 30 °C/min to 130 °C, heated at 5 °C/min to 250 °C, ramped to 300 °C at 10 °C/min, and maintained for 7 min. Helium (purity > 99.999%) was used as the carrier gas with a constant flow rate of 1.2 mL/min. The injection temperature was set to 270 °C and the injection volume was 1 µL. The injection mode was not split injection, and the purge valve was opened after 1 min.

The ion source was an electronic ionization source (70 eV, 280 °C), and the temperatures of the transfer line and the quadrupole were 250 °C and 180 °C, respectively. Solvent delay was set to 3 min; the ion monitoring mode was full scan; scanning ranged (m/z) from 45 to 550; the scan rate was 5 Hz. The total program duration was 42 min.

Mass calibration was required before sample acquisition, and the instrument was tuned at intervals to ensure stability.

2.6. Method Validation

The screening method of high-resolution mass spectrometry can be validated through screening detection limits (SDL), and the quantitative method can be validated through limit of quantitation (LOQ). The SDL, LOQ, linearity, recovery, and precision of this experiment were verified by SANTE/12682/2019 guidelines. SDL is the minimum concentration at which more than 95% of a series of concentration levels meets the detection requirements (20 additional experiments were conducted in parallel for each concentration). When the SDL and recovery were validated, all the target pesticides were spiked to the sample and the spiked samples were placed at room temperature for 30 min, then treated according to the above method. After the 10-point matrix matching calibration was constructed, its linearity was evaluated with the coefficient of determination (R^2). The recovery and precision were investigated in three different levels of spiked blank samples with 1-, 2-, and 10-times LOQ.

The matrix effect (ME) is the interference of other components in the matrix with the target compounds. The formula is:

$$ME (\%) = (bm - bs)/bs \times 100\% \qquad (1)$$

where bm is the slope of the matrix standard curve and bs is the slope of the solvent standard curve.

Based on previous studies, we established several hundred kinds of pesticide databases on gas and liquid high resolution mass spectrometry, respectively [25]. According to the recovery and precision, 237 pesticides were divided into pesticides suitable for GC or LC detection.

3. Results

3.1. Optimization of Hydration Volume

For the oily matrix, adding an appropriate amount of water for hydration during sample pretreatment was conducive to the softening of the matrix epidermis, making it easier for residual pesticides in the matrix to be extracted. This experiment explored the effect of different hydration volumes on the recovery of multiple pesticides. The experiment results show that the proportion of pesticides that met the recovery requirements (70–120%) under a non-hydration condition was 74.9%, which was less than under the conditions with water additions of 2 mL and 5 mL. Under the condition of a 2 mL water addition, the number of pesticides meeting the recovery requirements was the most numerous, accounting for 83.5%. As shown in Figure 1, the average recovery under the 2 mL condition was 88.3%, which was higher than that under the other two conditions. The results were in line with our expectations. The oil-water partition coefficient (logP) is an important parameter for the solubility of compounds, which is a simulated value based on the soil sorption coefficient normalized to organic carbon content (log Koc) [26]. The smaller the logP value, the better the water solubility of the compound. The effect of hydration volume on recovery with different logP was investigated, showing that hydration had a great impact on recovery with a low logP. The overall recovery of 54 pesticides with hydrophilic compounds (logP < 2.0) was low under a non-hydration condition, with the pesticides meeting the requirements accounting for 42.6%. When the hydration volume was 5 mL, the pores were opened due to the increase in the hydration volume, and multiple interferents in the matrix could be extracted together. The matrix promotion effect was enhanced, so that the overall recovery of pesticides with logP < 2.0 was higher than the recovery under the other two conditions. When the hydration volume was 2 mL, the pesticides that met the requirements of recovery were most numerous, accounting for 70.4%; therefore, 2 mL was finally selected as the optimal hydration volume.

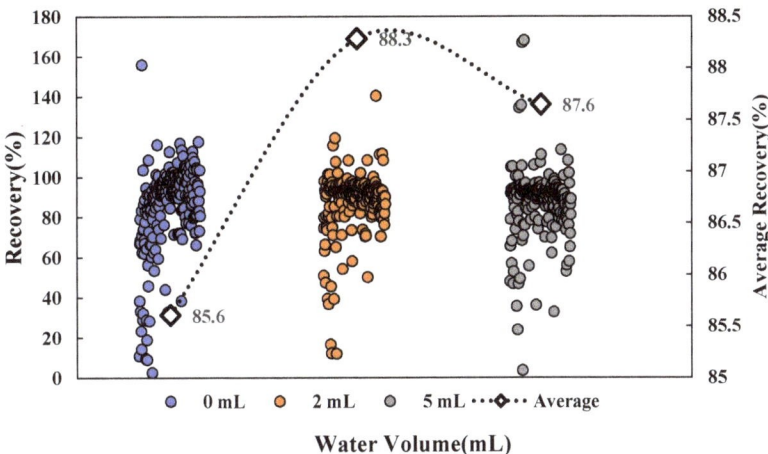

Figure 1. Effects of hydration volumes on pesticide recovery.

3.2. Optimization of Extraction Solvent Volume

The extraction of target compounds is a critical step in pesticide residue analysis. Mol et al. [27] tested a series of solvents for extraction and found that methanol usually extracts too many compounds in the matrix, and further matrix removal steps were required. Acetonitrile has low solubility in fat and a low matrix effect when extracting from complex matrices. Therefore, acetonitrile was selected as the extraction solvent of cottonseed hull in this experiment. Three different extraction volumes of 10 mL, 16 mL, and 20 mL (i.e., a hydration volume and extraction volume ratio of 1:5, 1:8, and 1:10) were compared to explore the effect of different extraction volumes on the recovery of pesticide residues. The results are shown in Figure 2. It was found that when the extract volume was 10 mL, 16 mL, and 20 mL, the proportion of pesticides meeting the recovery requirements was similar, at 81.0%, 80.7% and 81.3% respectively. However, at the spiked level, the volume of the extraction solution decreased, the pesticide concentration per unit volume increased, and more pesticide compounds had better peak shapes. In addition, a lower organic reagent amount was recommended from the perspective of green environmental protection, so the final extraction volume was 10 mL.

3.3. Optimization of Salting-Out Agent

The salting-out agents commonly used in pesticide residue screening were EN buffer salt (4 g $MgSO_4$, 1 g NaCl, 0.5 g disodium hydrogen citrate, and 1 g sodium citrate), the QuEChERS method for fruits and vegetables (4 g $MgSO_4$ and 1 g NaCl), and AOAC buffer salt (6 g $MgSO_4$ and 1.5 g NaAc). In this work, the effects of the above three salting-out agents on the recovery of pesticides were compared. As shown in Figure 3, although EN or AOAC salt forms a buffer system in the solution state, the results showed that the recovery using an $MgSO_4$ + NaCl combination best met the requirements, accounting for 78%. The reason for this result was that the volume of the extract from the QuEChERS method was relatively small. If the amount of extraction salt was too large, the heat emitted during water absorption destroys the structure of thermally unstable pesticides and affects their recovery. Therefore, 4 g MgSO$_4$ and 1 g NaCl with less salt consumption were finally selected as the salting-out agents.

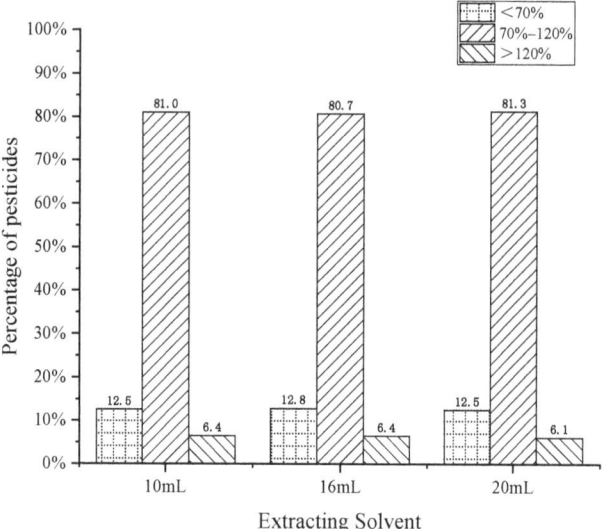

Figure 2. Effect of extraction solvent volume on pesticide recovery.

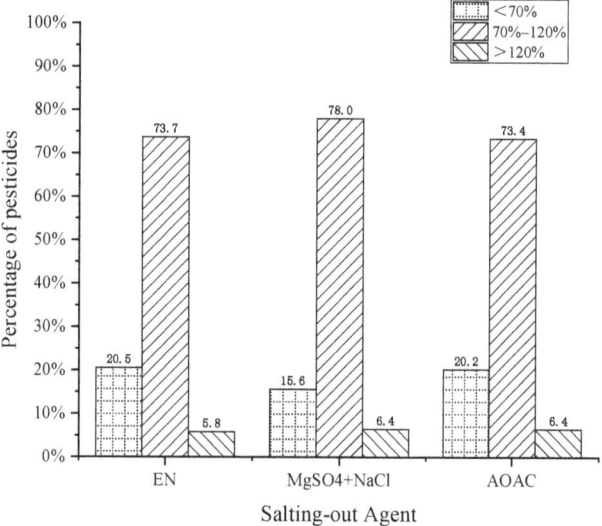

Figure 3. Effect of salting-out agents on pesticide recovery.

3.4. Optimization of Types and Amounts of Clean-Up Sorbents

A clean-up procedure was a key step in the pretreatment of the oily matrix. Its purpose was to effectively purify the analyzed matrix, and most of target pesticides had acceptable recovery, precision, and matrix effect [14]. Although acetonitrile had low liposolubility, which can slightly reduce the interference of a fat-soluble matrix on target compounds [15], in order to effectively reduce the influence of high-fat matrix co-extraction on the detection sensitivity of pesticides, as well as instrument loss, the clean-up procedure was necessary.

Theurillat established a d-SPE clean-up method containing 150 mg C18 and 150 mg PSA to determine 176 pesticide residues in fatty foods [12]. Therefore, this study was optimized on this basis.

In this work, the ability of MgSO$_4$ + PSA + C18 + Z-sep and MgSO$_4$ + PSA + C18 sorbents were compared. The structure of PSA had -NH$_2$, which can form a strong hydrogen bond with -COOH, so it was often used to adsorb polar compounds, such as fatty acids, lipids, and carbohydrates. C18 was often used to adsorb non-polar compounds, such as long-chain aliphatic compounds and sterols [8,25]. Z-sep was a new adsorbent, based on zirconia, which can be used for the adsorption of hydrophobic compounds in the fat matrix [28]. It was seen that the bottom of the purification tube after Z-sep purification was dark yellow, while the sample without Z-sep purification was light yellow, indicating that Z-sep had an obvious effect on degreasing.

In order to further verify the ability of sorbents, the spiked experiments were carried out. As shown in Figure 4, A was the sorbent combination of MgSO$_4$ + PSA + C18 + Z-sep, and B was the sorbent combination of MgSO$_4$ + PSA + C18. As a result, the sorbent combination without Z-sep accounted for more pesticides that meet the requirements, reaching 81.04%. The reason for this result was that Z-sep adsorbs some target pesticides while removing lipids. According to the Lewis theory, the affinities of Z-sep on the analyte containing different substituent characteristics can be sorted in the following order: chloride < formate < acetate < sulphate < citrate < fluoride < phosphate < hydroxide [25]. In this work, a variety of pesticides, such as trinexapac-ethyl, abamectin containing -OH, fenamiphos sulfoxide containing phosphate, and sulfoxaflor containing sulphate, had substituents with a strong affinity to Z-sep. Therefore, the recovery of sorbent combinations with Z-sep was significantly lower than that without Z-sep. Although Z-sep was more efficient in removing lipid compounds, the sorbent combination of MgSO$_4$ + PSA + C18 was finally selected as the purification filler in this work, from the perspective of method versatility.

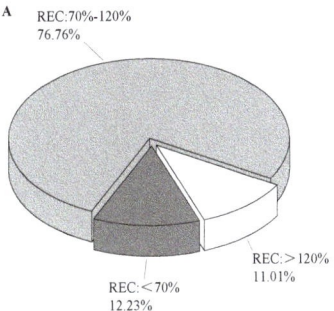

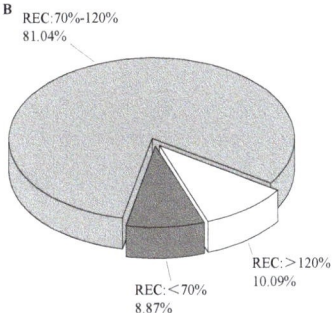

Figure 4. Effect of clean-up sorbents on pesticide recovery. (**A**) MgSO$_4$ + PSA + C18 + Z-sep; (**B**) MgSO$_4$ + PSA + C18.

The amount of PSA and C18 was also optimized. The effects of PSA (50–150 mg) and C18 (100–300 mg) on the recovery of various pesticides were optimized by controlling other variables. The results showed that when the amount of PSA was 100 mg, the greatest number of pesticides with satisfactory recovery was obtained, accounting for 73.7%. With the increase in PSA amount, the recovery of organic nitrogen pesticides, such as propanil and fenbuconazole, and carbamate pesticides, such as aldicarb-sulfone and thiophanate-methyl, gradually decreased. When the amount of C18 was 100 mg, the proportion of pesticides that met satisfactory recovery was 82.0%. With an increase in the C18 amount, the recovery of various organic nitrogen pesticides obviously decreased, especially the chlorides with a benzene ring structure, such as monolinuron, novaluron, propanil, and pretilachlor. Therefore, 100 mg PSA and 100 mg C18 were finally selected as the optimal amounts of clean-up sorbents.

3.5. Evaluation of Matrix Effect

Analysis of pesticide residues in the oil matrix may be adversely affected by the matrix effect. The main result of the matrix effect is to increase or decrease the analyte signal when the same analyte exists in the solvent [29]. The methods for eliminating or reducing the matrix effect include: (1) optimizing the sample preparation method and reducing co-extraction; (2) changing the chromatographic mass spectrometry conditions; (3) diluting the samples; and (4) using matrix-matched standards or an additional standard method [30]. In this work, the purifying agent was optimized, and the matrix-matched standard was used to reduce the interference of the matrix effect on target compounds. The matrix effect distribution of 237 pesticides is shown in Figure 5. Among the 237 pesticides investigated in cottonseed hull samples, the proportion of pesticides with a negative matrix effect accounted for 81.4%, indicating that the substrate had a suppression effect on the tested pesticides as a whole. The matrix effect can be divided into three categories: no matrix effect ($|ME| \leq 20\%$); a weak matrix effect ($20\% < |ME| < 50\%$); and a strong matrix effect ($|ME| \geq 50\%$). In this work, only 8% of the pesticides in the cottonseed hull matrix showed a strong matrix effect; the weak matrix effect and no matrix effect accounted for 13.1% and 78.9%, respectively, indicating that this research method had a strong anti-matrix interference ability.

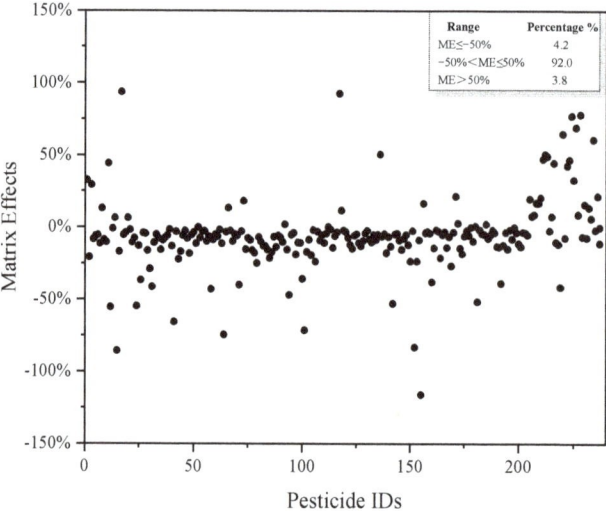

Figure 5. Matrix effect distribution of 237 pesticides.

3.6. Method Validation and Method Performance

3.6.1. SDL, LOQ, and Standard Curve

The method validation was carried out under the optimal sample preparation procedure, and the results are shown in Table 1. The typical extraction ion chromatograms of GC-Q TOF/MS and LC-Q TOF/MS are shown in Figures 6 and 7, respectively. The SDLs were in the range of 0.2–20 µg/kg, of which 224 pesticides (accounting for 94.5%) were in the range of 0.2–5 µg/kg. The LOQs were in the range of 0.2–20 µg/kg; 215 pesticides (accounting for 90.7%) had an LOQ range of 0.2–5 µg/kg. Shinde developed and verified 222 and 220 multi-pesticides residue analysis methods in sesame seeds, using LC-MS/MS and GC-MS/MS, respectively, and most pesticides offered an LOQ of 10 µg/kg for most compounds [16]. Kuzukiran et al. developed an SPE sample preparation method, combined with GC-MS, GC-MS/MS and LC-MS/MS, to analyze the residues of 322 organic pollutants in bats [31]. The LOQ of the method was in the range of 0.27–19.26 µg/kg, which was similar to that in our work; however, they paid more attention to environmental pollutants. This indicated that this method had high sensitivity in the detection of pesticide residues in cottonseed hull matrix. It is noteworthy that due to the large number of pesticides spiked, the retention time of some pesticides may overlap or be very close; for example, the RTs of Chloridazon and Mevinphos were 3.62 min. However, the excellent resolution of high-resolution mass spectrometry was sufficient to separate compounds that had a similar RT but a different mass (the quantitative ion mass of Chloridazon and that of Mevinphos were 222.04287 and 225.05230, respectively).

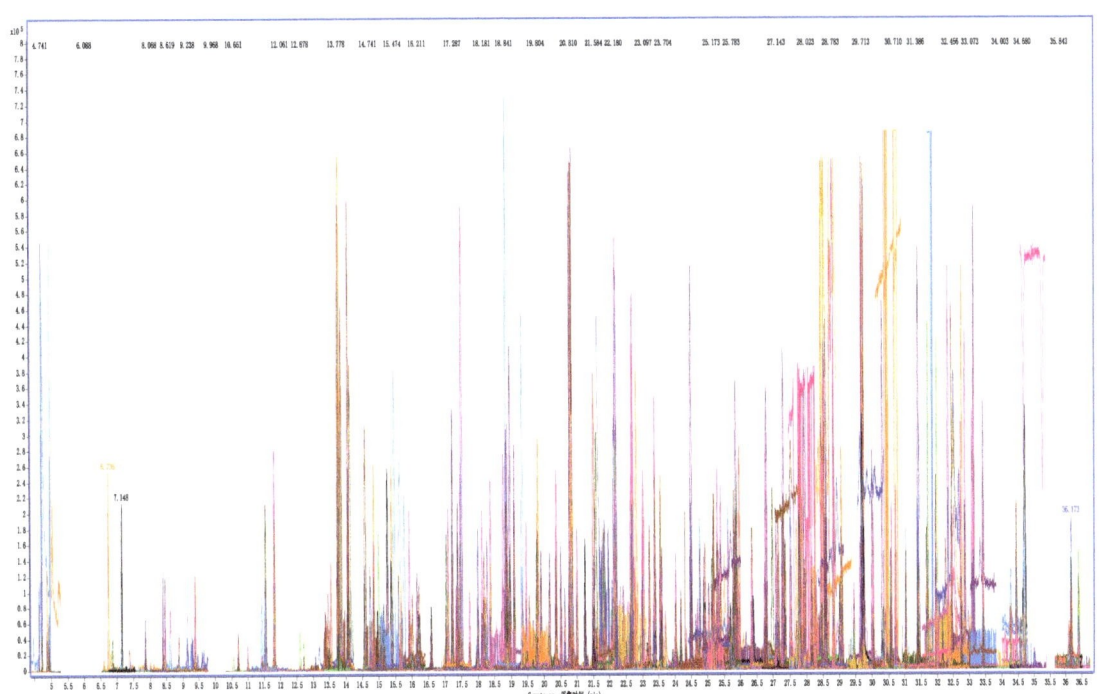

Figure 6. Overlay extraction ion chromatograms of GC-Q TOF/MS of cottonseed hull sample at spiking level of 200 µg/kg.

Table 1. Compound information, screening detection limits (SDLs), limit of quantification (LOQ), linear range, R^2, recovery, and RSD of 237 pesticides (n = 6).

No	Compound	Formula	RT (Min)	Quantitative Ion	Qualitative Ion	R^2	Linearity (ng/g)	SDL (ng/g)	LOQ (ng/g)	1-LOQ REC (%)	1-LOQ RSD (%)	2-LOQ REC (%)	2-LOQ RSD (%)	10-LOQ REC (%)	10-LOQ RSD (%)	Detecting Instrument
1	1-(2-chloro-4-(4-chlorophenoxy)phenyl)-2-(1H-1,2,4-triazol-1-yl)ethanol	C16H13Cl2N3O2	10.16	350.04580	70.03997	0.9995	10–200	10	10	81.3	2.9	71.5	7.8	76.9	11.5	LC
2	1-(2-Chloro-pyridin-5-yl-methyl)-2-imino-imidazolidine hydrochloride	C9H11ClN4	2.28	211.07450	90.03383	0.9992	1–200	1	1	71.4	1.4	58.7	4.7	54.4	1.1	LC
3	1-methyl-3-(tetrahydro-3-furylmethyl) urea	C7H14N2O2	1.87	159.11280	58.02874	0.9986	2–200	2	2	84.1	17.2	78.2	11.8	87.3	7.9	LC
4	2,4-D butylate	C12H14O3Cl2	19.45	185.00000	276.03146	0.9992	1–200	1	1	71.7	14.3	90.8	5.3	77.6	4.6	GC
5	2-(Trifluoromethyl)-1-methyl-1H-pyrazole-4-carboxamide	C6H6F3N3O	2.63	194.05360	134.03488	0.9911	20–200	5	20	104.7	8.7	105.4	12.8	79.6	3.2	LC
6	5-hydroxy Imidacloprid	C9H10ClN5O3	3.05	272.05450	225.05377	0.9976	10–200	5	10	72.0	7.8	78.2	8.5	70.2	7.8	LC
7	Acetamiprid	C10H11ClN4	3.93	223.07450	126.01051	0.9938	1–200	1	1	93.5	11.7	110.3	11.0	89.6	7.2	LC
8	Acetamiprid-N-desmethyl	C9H9ClN4	3.57	209.05890	126.01051	0.9950	1–200	1	1	113.2	9.4	105.6	8.8	96.3	6.4	LC
9	Acetochlor	C14H20ClNO2	12.57	270.12553	133.08861	0.9992	1–200	1	1	115.3	13.8	99.2	10.5	86.4	3.1	LC
10	Alachlor	C14H20ClNO2	12.45	270.12553	238.09932	0.9995	1–200	1	1	99.2	15.1	70.1	7.9	84.8	4.1	LC
11	Aldicarb-sulfone	C7H14N2O4S	2.63	223.07470	62.98991	0.9987	10–200	10	10	73.7	7.5	86.5	8.3	84.2	2.7	LC
12	Aldrin	C12H8Cl6	19.52	262.85641	264.85352	0.9994	1–200	1	1	79.7	4.5	80.8	7.4	66.3	3.5	GC
13	Allidochlor	C8H12ClNO	4.94	174.06800	98.09643	0.9995	0.2–200	0.2	0.2	93.1	19.9	70.7	15.3	83.8	8.1	LC
14	Alpha-HCH	C6H6Cl6	16.14	182.93437	180.93732	0.9904	1–200	1	1	99.6	10.4	71.6	8.2	87.7	8.2	GC
15	Ametryn	C9H17N5S	6.70	228.12774	186.08080	0.9997	1–200	1	1	82.6	2.5	91.2	1.1	79.3	1.9	LC
16	Atrazine	C8H14ClN5	6.38	216.10105	174.05409	0.9999	1–200	1	1	82.7	3.9	93.1	3.1	78.1	1.5	LC
17	Atrazine D5 (Ethylamino D5)	C8H9D5ClN5	6.47	221.13310	69.03060	0.9965	1–200	1	1	97.7	3.5	100.0	0.9	98.6	1.4	LC
18	Avermectin B1a	C48H72O14	18.66	895.48140	751.40521	0.9983	2–200	1	2	90.9	11.6	83.7	3.3	110.0	6.5	LC
19	Azoxystrobin	C22H17N3O5	11.07	404.12410	329.07950	0.9996	1–200	1	1	94.8	8.1	93.9	2.8	82.2	1.3	LC
20	Benalaxyl	C20H23NO3	14.04	326.17507	91.05423	0.9999	1–200	1	1	91.6	5.6	96.9	3.2	83.2	2.2	LC
21	Bendiocarb	C11H13NO4	5.74	224.09173	81.03349	1.0000	2–200	1	2	85.9	8.2	84.5	17.1	120.0	8.7	LC
22	Benfluralin	C13H16F3N3O4	15.37	292.05396	264.02267	1.0000	50–200	2	20	96.7	5.2	73.7	2.4	73.7	7.9	GC
23	Benfuracarb	C20H30N2O5S	17.27	411.19482	102.00081	0.9999	1–200	1	1	64.2	5.0	71.9	9.4	44.3	2.6	LC
24	Benzovindiflupyr	C18H15Cl2F2N3O	14.33	398.06400	159.03644	0.9999	1–200	1	1	95.9	5.0	98.1	2.2	83.0	1.8	LC
25	beta-Endosulfan	C9H6Cl6O3S	27.04	236.84077	242.90135	0.9950	1–200	1	1	103.9	14.0	91.7	5.7	70.3	9.0	GC
26	Beta-HCH	C6H6Cl6	20.76	182.93437	180.93732	0.9964	1–200	1	1	72.3	17.2	70.5	15.6	74.2	7.7	GC
27	Bifenazate	C17H20N2O3	12.18	301.15467	198.09134	0.9999	1–200	1	1	89.5	18.7	83.4	12.2	88.0	6.3	LC

Table 1. Cont.

No	Compound	Formula	RT (Min)	Quantitative Ion	Qualitative Ion	R^2	Linearity (ng/g)	SDL (ng/g)	LOQ (ng/g)	1-LOQ REC (%)	1-LOQ RSD (%)	2-LOQ REC (%)	2-LOQ RSD (%)	10-LOQ REC (%)	10-LOQ RSD (%)	Detecting Instrument
28	Bifenthrin	C23H22ClF3O2	28.81	181.10118	166.07770	0.9939	10–200	10	10	106.6	16.2	45.7	2.6	82.8	2.0	GC
29	Bioresmethrin	C22H26O3	19.09	339.19550	143.08553	0.9978	20–200	20	20	84.9	17.6	71.6	6.6	77.7	6.0	LC
30	Bitertanol	C20H23N3O2	12.68	338.18630	70.03997	0.9928	5–200	5	5	77.1	10.6	79.2	10.2	75.8	14.2	LC
31	Boscalid	C18H12Cl2N2O	11.18	343.03994	271.08658	0.9997	2–200	1	2	100.9	9.9	91.1	10.5	88.0	7.9	LC
32	Bromobutide	C15H22BrNO	13.75	312.09575	119.08553	0.9998	1–200	1	1	91.3	18.0	92.3	14.9	70.3	7.0	LC
33	Bromophos-methyl	C8H8BrCl2O3PS	21.82	330.87753	328.87982	0.9941	1–200	1	1	75.4	15.6	81.2	10.8	78.4	14.3	GC
34	Bromopropylate	C17H16Br2O3	29.69	340.89948	342.89755	0.9989	1–200	1	1	87.1	10.5	90.6	6.9	76.6	7.2	GC
35	Bupirimate	C13H24N4O3S	12.61	317.16419	44.04948	0.9998	1–200	1	1	95.0	8.2	95.3	2.6	82.3	1.0	LC
36	Buprofezin	C16H23N3OS	17.38	306.16346	57.06988	0.9975	1–200	1	1	101.4	3.6	101.9	17.7	78.7	3.1	LC
37	Butachlor	C17H26ClNO2	17.47	312.17250	57.06988	0.9987	1–200	1	1	115.3	19.9	95.0	19.9	80.4	6.5	LC
38	Butamifos	C13H21N2O4PS	16.45	333.10350	95.96675	0.9981	1–200	1	1	86.8	14.4	110.6	9.5	74.7	6.5	LC
39	Butylate	C11H23NOS	16.60	218.15731	57.06988	0.9992	10–200	5	10	61.4	10.4	82.4	6.5	70.4	15.9	LC
40	Cadusafos	C10H23O2PS2	14.61	271.09498	96.95076	0.9998	1–200	1	1	73.3	9.4	86.1	9.1	77.3	1.8	LC
41	Carbaryl	C12H11NO2	6.21	202.08626	127.05423	0.9927	10–200	10	10	102.3	17.1	112.5	16.9	107.1	17.2	LC
42	Carbendazim	C9H9N3O2	2.67	192.07675	160.05054	0.9999	1–200	1	1	111.8	19.1	79.8	9.6	71.5	12.1	LC
43	Carbofuran	C12H15NO3	5.80	222.11247	123.04406	0.9944	1–200	1	1	107.5	3.6	91.3	6.1	112.5	2.4	LC
44	Carbofuran-3-Hydroxy	C12H15NO4	3.55	238.10738	107.04914	0.9997	0.2–200	0.2	0.2	82.7	17.6	70.6	11.0	86.9	9.3	LC
45	Carbosulfan	C20H32N2O3S	19.82	381.22064	76.02155	0.9994	2–200	2	2	79.8	18.9	34.2	11.0	51.7	12.0	LC
46	Carfentrazone-ethyl	C15H14Cl2F3N3O3	14.18	412.04350	345.99561	0.9998	1–200	1	1	115.9	14.3	85.8	7.5	84.7	4.0	LC
47	Chlorantraniliprole	C18H14BrCl2N5O2	8.23	481.97807	283.92160	1.0000	1–200	1	1	93.0	16.9	99.8	10.3	81.1	4.9	LC
48	Chlorfenapyr	C15H11BrClF3N2O	27.57	363.94073	361.94278	0.9913	1–200	1	1	85.1	6.6	109.8	17.5	112.4	13.6	GC
49	Chlorfenvinphos	C12H14Cl3O4P	13.67	358.97681	98.98434	0.9999	1–200	1	1	89.8	10.9	103.8	9.8	86.5	4.2	LC
50	Chloridazon	C10H8ClN3O	3.62	222.04287	77.03857	0.9977	1–200	1	1	77.7	6.2	84.4	4.1	79.2	3.4	LC
51	Chlormequat	C5H12ClN	0.70	122.07310	58.06512	0.9983	1–200	1	1	90.8	14.7	93.6	4.7	113.3	8.4	LC
52	Chloroneb	C8H8Cl2O2	11.81	190.96611	192.96324	0.9945	2–200	2	2	127.8	19.3	54.8	13.8	69.5	8.8	GC
53	Chlorotoluron	C10H13ClN2O	6.10	213.07892	72.04488	0.9998	1–200	1	1	96.1	6.1	102.1	5.0	82.2	2.2	LC
54	Chlorpropham	C10H12ClNO2	15.92	127.01833	213.05511	0.9989	5–200	5	5	138.2	19.7	88.1	15.8	84.0	8.3	GC
55	Chlorpyrifos	C9H11Cl3NO3PS	17.72	349.93356	96.95076	0.9998	5–200	1	5	100.9	19.3	84.9	8.9	85.1	5.3	LC
56	Chlorpyrifos-methyl	C7H7Cl3NO3PS	19.32	285.92557	287.92316	0.9949	1–200	1	1	93.6	7.9	79.4	10.6	78.7	13.1	GC
57	Cis-Chlordane (alpha)	C10H6Cl8	23.58	372.82544	374.82251	0.9996	1–200	1	1	85.7	7.4	83.5	6.9	70.5	6.5	GC
58	Clodinafop-propargyl	C17H13ClFNO4	15.05	350.05899	91.05423	0.9999	1–200	1	1	109.5	7.4	94.4	4.1	83.1	3.0	LC
59	Clofentezine	C14H8Cl2N4	15.32	303.01988	102.03383	0.9997	5–200	5	5	81.2	17.4	72.2	12.4	104.2	7.5	LC
60	Clomazone	C12H14ClNO2	7.91	240.07858	125.01525	0.9999	1–200	1	1	112.3	18.4	99.8	12.7	78.5	3.6	LC
61	Clothianidin	C6H8ClN5O2S	3.50	250.01600	131.96692	0.9995	2–200	1	2	101.9	19.4	103.9	15.1	89.4	9.9	LC
62	Cyanazine	C9H13ClN6	5.16	241.09630	214.08540	0.9998	1–200	1	1	78.9	11.9	89.0	8.2	85.9	13.9	LC
63	Cyanofenphos	C15H14NO2PS	29.06	156.98715	169.04129	0.9996	1–200	1	1	77.3	14.9	100.8	13.9	81.2	6.1	GC

Table 1. Cont.

No	Compound	Formula	RT (Min)	Quantitative Ion	Qualitative Ion	R^2	Linearity (ng/g)	SDL (ng/g)	LOQ (ng/g)	1-LOQ REC (%)	1-LOQ RSD (%)	2-LOQ REC (%)	2-LOQ RSD (%)	10-LOQ REC (%)	10-LOQ RSD (%)	Detecting Instrument
64	Cycloate	C11H21NOS	15.35	216.14166	55.05423	0.9998	2–200	2	2	117.4	15.3	85.0	16.2	79.6	5.7	LC
65	Cycloxydim	C17H27NO3S	16.22	326.17844	107.04914	0.9996	1–200	1	1	62.6	11.6	100.4	5.7	74.1	4.2	LC
66	Cyprodinil	C14H15N3	22.15	224.11823	225.12605	0.9999	1–200	1	1	92.3	15.9	89.1	4.1	75.3	5.9	GC
67	Cyromazine	C6H10N6	0.75	167.10400	85.05087	0.9927	5–200	1	5	58.2	6.0	56.2	10.1	51.5	5.8	LC
68	Delta-HCH	C6H6Cl6	21.60	180.93732	182.93437	0.9943	2–200	2	2	158.1	19.5	162.4	19.8	181.8	19.4	GC
69	Desmetryn	C8H15N5S	5.21	214.11209	172.06514	0.9994	1–200	1	1	84.7	3.7	93.5	0.7	78.6	1.9	LC
70	Diallate	C10H17Cl2NOS	16.66	270.04810	86.06004	0.9992	10–200	5	10	70.6	19.0	107.5	9.7	85.9	19.9	LC
71	Diazinon	C12H21N2O3PS	14.97	305.10833	96.95076	0.9998	1–200	1	1	89.5	3.6	87.7	3.4	78.5	1.5	LC
72	Dichlofenthion	C10H13Cl2O3PS	18.86	279.00061	222.93800	0.9938	1–200	1	1	76.6	17.7	82.0	10.5	73.0	7.0	LC
73	Dichlorvos	C4H7Cl2O4P	7.85	184.97650	109.00491	0.9954	10–200	1	10	116.2	12.6	108.9	17.6	78.1	16.0	GC
74	Dicloran	C6H4Cl2N2O2	18.20	205.96443	207.96156	0.9946	2–200	2	2	89.3	15.9	89.7	14.5	86.0	8.9	GC
75	Difenoconazole	C19H17Cl2N3O3	14.63	406.07200	251.00250	0.9998	1–200	1	1	77.3	4.8	94.0	2.8	79.6	2.2	LC
76	Diflubenzuron	C14H9ClF2N2O2	12.11	311.03934	141.01465	0.9938	10–200	10	10	64.8	5.6	83.7	10.3	92.2	10.0	LC
77	Dimethenamid	C12H18ClNO2S	9.58	276.08195	244.05574	0.9997	1–200	1	1	90.6	3.6	84.8	5.5	82.8	4.6	LC
78	Dimethoate	C5H12NO3PS2	3.78	230.00690	198.96469	0.9941	2–200	1	2	70.4	10.0	89.8	9.0	84.7	6.0	LC
79	Dimethylvinphos (Z)	C10H10Cl3O4P	10.47	330.94550	127.01547	0.9999	1–200	1	1	92.1	16.9	80.7	10.3	108.7	7.9	LC
80	Diniconazole	C15H17Cl2N3O	12.97	326.08210	70.03997	1.0000	2–200	1	2	89.1	4.0	84.2	7.1	85.5	5.5	LC
81	Dinotefuran	C7H14N4O3	2.31	203.11387	58.05255	0.9994	20–200	20	20	80.0	5.6	86.8	4.5	77.1	3.3	LC
82	Dioxabenzofos	C8H9O3PS	10.47	217.00830	77.03857	1.0000	2–200	1	2	97.3	5.0	88.0	5.7	87.0	2.9	LC
83	Dipropetryn	C11H21N5S	11.46	256.15904	102.01205	0.9999	1–200	1	1	74.3	5.0	94.6	4.2	76.3	3.1	LC
84	Diuron	C9H10Cl2N2O	6.63	233.02429	72.04488	0.9989	1–200	1	1	105.7	17.6	124.8	11.1	70.6	5.3	LC
85	Edifenphos	C14H15O2PS2	13.46	311.03238	109.01065	0.9998	1–200	1	1	92.6	4.6	93.3	3.6	81.5	0.8	LC
86	Emamectin B1a	C49H75NO13	16.88	886.53112	158.11755	0.9996	1–200	1	1	83.1	12.3	82.3	6.6	75.2	10.3	LC
87	Endosulfan-sulfate	C9H6Cl6O4S	29.05	271.80963	273.80667	0.9999	1–200	1	1	61.7	7.1	61.1	6.0	51.4	2.7	GC
88	Ethalfluralin	C13H14F3N3O4	14.96	276.05905	316.09036	0.9981	1–200	1	1	93.0	9.6	106.1	16.4	73.9	9.5	GC
89	Ethion	C9H22O4P2S4	17.95	384.99489	199.00108	1.0000	1–200	1	1	119.9	8.3	89.1	16.0	78.1	3.3	LC
90	Ethoprophos	C8H19O2PS2	10.86	243.06368	96.95076	0.9998	1–200	1	1	90.3	11.8	87.3	6.6	78.7	1.5	LC
91	Etrimfos	C10H17N2O4PS	14.56	293.07194	124.98206	0.9999	1–200	1	1	76.9	4.5	92.3	4.6	81.6	3.0	LC
92	Fenamidone	C17H17N3OS	30.72	268.09030	238.11006	0.9994	1–200	1	1	77.7	11.7	89.6	17.0	87.1	6.1	GC
93	Fenamiphos	C13H22NO3PS	10.46	304.11308	201.98480	0.9998	1–200	1	1	83.6	2.6	96.2	2.9	84.2	0.9	LC
94	Fenamiphos-sulfone	C13H22NO5PS	5.59	336.10291	266.02466	0.9999	1–200	1	1	94.6	18.1	91.1	4.4	83.4	2.0	LC
95	Fenamiphos-sulfoxide	C13H22NO4PS	4.61	320.10799	108.05727	0.9999	1–200	1	1	90.3	5.0	100.2	6.5	83.9	1.1	LC
96	Fenarimol	C17H12Cl2N2O	10.59	331.03994	81.04472	0.9998	2–200	1	2	102.8	11.0	90.7	9.7	76.6	6.8	LC
97	Fenbuconazole	C19H17ClN4	12.38	337.12150	70.03997	0.9999	1–200	1	1	116.4	11.7	74.0	19.3	75.1	7.0	LC
98	Fenchlorphos	C8H8Cl3O3PS	19.80	284.93033	286.92749	0.9968	1–200	1	1	86.2	7.0	102.3	6.6	75.3	9.4	GC
99	Fenobucarb	C12H17NO2	8.80	208.13321	77.03857	0.9982	5–200	5	5	87.9	12.4	104.7	8.0	84.0	2.7	LC

Table 1. Cont.

No	Compound	Formula	RT (Min)	Quantitative Ion	Qualitative Ion	R^2	Linearity (ng/g)	SDL (ng/g)	LOQ (ng/g)	1-LOQ REC (%)	1-LOQ RSD (%)	2-LOQ REC (%)	2-LOQ RSD (%)	10-LOQ REC (%)	10-LOQ RSD (%)	Detecting Instrument
100	Fenpropimorph	C20H33NO	18.52	128.10699	129.11012	0.9948	5–200	5	5	66.6	19.4	70.7	16.3	76.9	3.2	GC
101	Fensulfothion	C11H17O4PS2	7.42	309.03786	140.02904	0.9996	1–200	1	1	87.9	2.9	94.5	1.5	84.0	1.2	LC
102	Fenthion-sulfoxide	C10H15O4PS2	6.02	295.02221	109.00491	0.9998	1–200	1	1	87.3	3.7	93.5	3.7	84.1	1.1	LC
103	Fipronil	C12H4Cl2F6N4OS	28.19	366.94296	368.94003	0.9970	1–200	1	1	80.1	19.1	75.7	11.4	72.0	7.6	GC
104	Fipronil Desulfinyl	C12H4Cl2F6N4	25.54	332.99609	387.97116	0.9951	2–200	2	2	107.4	12.3	67.2	18.7	117.8	19.5	GC
105	Fipronil-sulfide	C12H4Cl2F6N4S	27.81	350.94803	352.94510	0.9999	1–200	1	1	75.5	2.9	75.5	3.0	71.2	2.0	GC
106	Fluacrypyrim	C20H21F3N2O5	16.67	427.14753	145.06479	0.9992	1–200	1	1	109.3	15.5	111.5	6.5	77.6	6.8	LC
107	Fluazifop-butyl	C19H20F3NO4	17.62	384.14172	91.05423	0.9999	1–200	1	1	76.7	5.6	91.6	3.3	80.5	3.0	LC
108	Flubendiamide	C23H22F7IN2O4S	14.52	705.01250	530.97986	0.9999	1–200	1	1	88.5	3.1	93.2	3.4	83.2	1.7	LC
109	Flumiclorac-pentyl	C21H23ClFNO5	17.47	441.15930	308.04843	0.9973	2–200	1	2	26.4	19.5	75.0	19.5	79.0	18.1	LC
110	Fluopicolide	C14H8Cl3F3N2O	11.85	382.97271	172.95555	0.9999	1–200	1	1	90.7	9.5	95.1	6.7	83.8	2.4	LC
111	Fluquinconazole	C16H8Cl2FN5O	11.40	376.01630	306.98358	0.9996	5–200	5	5	82.4	6.4	88.9	6.3	94.9	4.3	LC
112	Fluridone	C19H14F3NO	9.19	330.11003	309.09598	0.9989	1–200	1	1	92.3	8.5	94.2	1.4	84.3	2.0	LC
113	Flusilazole	C16H15F2N3Si	12.36	316.10761	165.06967	0.9997	1–200	1	1	82.1	6.5	97.4	9.5	79.8	2.3	LC
114	Flutriafol	C16H13F2N3O	6.40	302.10994	70.03997	0.9996	1–200	1	1	89.3	9.1	96.2	5.3	76.0	3.2	LC
115	Fluxapyroxad	C18H12F5N3O	11.39	382.09730	342.08487	1.0000	1–200	1	1	93.0	7.8	94.4	4.5	84.2	3.4	LC
116	Fonofos	C10H15OPS2	15.23	247.03747	80.95585	0.9976	5–200	1	5	72.0	1.5	100.3	16.0	105.8	4.3	LC
117	Fosthiazate	C9H18NO3PS2	6.37	284.05385	104.01646	0.9998	1–200	1	1	96.0	11.1	94.6	4.3	87.1	2.2	LC
118	Furathiocarb	C18H26N2O5S	17.26	383.16352	195.04742	0.9999	1–200	1	1	74.3	6.7	82.6	4.4	64.9	1.2	LC
119	Haloxyfop	C15H11ClF3NO4	23.54	316.03467	375.04797	0.9997	20–200	1	20	96.0	4.8	77.0	6.3	73.3	6.9	GC
120	Haloxyfop-2-ethoxyethyl	C19H19ClF3NO5	17.06	434.09766	91.05423	0.9983	1–200	1	1	92.3	2.4	110.3	3.5	83.7	3.2	LC
121	Haloxyfop-methyl	C16H13ClF3NO4	16.23	376.05460	272.00845	0.9985	1–200	1	1	91.7	11.5	90.2	5.9	83.0	2.6	LC
122	Heptachlor	C10H5Cl7	18.48	271.80963	273.80667	0.9979	1–200	1	1	106.4	18.9	76.3	6.1	74.9	10.0	GC
123	Hexachlorobenzene	C6Cl6	14.03	283.80963	285.80670	0.9918	1–200	1	1	61.6	2.1	60.1	4.1	54.3	6.4	GC
124	Hexaconazole	C14H17Cl2N3O	12.19	314.08250	70.03997	0.9996	2–200	1	2	98.4	11.4	85.7	9.5	97.2	10.8	LC
125	Hexythiazox	C17H21ClN2O2S	17.70	353.10850	168.05696	0.9992	2–200	1	2	70.6	10.4	120.0	9.3	82.0	4.1	LC
126	Imazalil	C14H14Cl2N2O	25.76	172.95555	215.00250	0.9998	1–200	1	1	92.1	6.8	73.2	7.6	84.1	3.9	GC
127	Imazapyr	C13H15N3O3	3.07	262.11862	69.06988	0.9998	5–200	5	5	24.8	2.0	23.1	11.0	25.3	6.7	LC
128	Imidacloprid	C9H10ClN5O2	3.68	256.05958	209.05885	0.9967	1–200	1	1	117.5	0.2	181.6	6.6	87.1	12.1	LC
129	Imidacloprid-Olefin	C9H8ClN5O2	3.07	254.04390	171.06653	0.9997	10–200	5	10	94.8	13.7	91.1	7.3	78.1	9.6	LC
130	Iprobenfos	C13H21O3PS	12.36	289.10218	91.05423	0.9994	2–200	1	2	104.1	14.5	100.1	15.0	101.6	5.7	LC
131	Iprovalicarb	C18H28N2O3	10.44	321.21727	119.08553	0.9999	1–200	1	1	111.8	13.5	106.4	9.2	89.1	2.3	LC
132	Isazofos	C9H17ClN3O3PS	13.62	314.04895	119.99574	0.9998	1–200	1	1	86.0	4.6	91.2	3.4	83.4	2.0	LC
133	Isofenphos	C15H24NO4PS	16.48	346.12364	121.02872	0.9996	5–200	2	5	76.4	15.6	80.4	17.8	108.1	16.7	LC
134	Isoproturon	C12H18N2O	6.66	207.14919	72.04439	0.9990	0.5–200	0.5	0.5	96.2	7.9	89.2	1.7	84.0	2.2	LC
135	Isopyrazam	C20H23F2N3O	15.58	360.18950	320.17575	0.9998	1–200	1	1	86.9	5.8	96.8	4.1	79.9	1.2	LC

Table 1. Cont.

No	Compound	Formula	RT (Min)	Quantitative Ion	Qualitative Ion	R²	Linearity (ng/g)	SDL (ng/g)	LOQ (ng/g)	1-LOQ REC (%)	1-LOQ RSD (%)	2-LOQ REC (%)	2-LOQ RSD (%)	10-LOQ REC (%)	10-LOQ RSD (%)	Detecting Instrument
136	Kresoxim-methyl	C18H19NO4	14.26	314.13868	116.04948	0.9998	5–200	2	5	72.0	15.1	79.3	12.3	89.8	4.7	LC
137	Lactofen	C19H15ClF3NO7	17.70	479.08210	343.99319	0.9972	20–200	20	20	90.5	5.1	111.6	17.1	77.7	12.5	LC
138	Lindane	C6H6Cl6	17.74	180.93732	182.93437	0.9989	1–200	1	1	62.9	16.7	116.5	18.3	110.8	7.9	GC
139	Linuron	C9H10Cl2N2O2	9.10	249.01921	132.96063	0.9990	5–200	2	5	73.1	10.3	84.4	8.9	94.2	3.2	LC
140	Malaoxon	C10H19O7PS	5.72	315.06619	99.00767	0.9998	1–200	1	1	74.4	17.4	93.9	12.3	89.4	10.2	LC
141	Malathion	C10H19O6PS2	12.53	331.04334	99.00767	0.9983	1–200	1	1	82.9	11.0	83.7	10.3	77.8	3.4	LC
142	Mepanipyrim	C14H13N3	24.48	222.10257	223.11040	0.9998	5–200	1	5	84.5	9.9	77.6	8.7	79.4	4.8	GC
143	Metaflumizone	C24H16F6N4O2	17.39	507.12502	178.04628	0.9973	10–200	2	10	80.3	4.6	86.7	16.9	82.8	8.0	LC
144	Metalaxyl	C15H21NO4	6.70	280.15433	45.03349	0.9993	1–200	1	1	95.0	8.6	98.4	2.2	81.8	1.1	LC
145	Metconazole	C17H22ClN3O	12.66	320.15221	70.03997	0.9998	2–200	1	2	80.5	8.7	86.5	5.8	86.1	6.8	LC
146	Methiocarb	C11H15NO2S	8.73	226.08960	121.06479	0.9939	20–200	5	20	84.1	15.4	86.1	16.2	85.3	5.1	LC
147	Methiocarb-sulfoxide	C11H15NO3S	3.42	242.08454	122.07262	0.9980	1–200	1	1	101.1	14.8	87.6	6.8	84.0	4.0	LC
148	Metolachlor	C15H22ClNO2	12.32	284.14118	252.11497	0.9999	1–200	1	1	97.1	9.1	105.6	3.2	84.0	1.4	LC
149	Metrafenone	C19H21BrO5	16.24	409.06451	209.08084	0.9998	1–200	1	1	91.0	4.6	92.7	3.6	79.1	1.9	LC
150	Metribuzin	C8H14N4OS	5.26	215.09611	49.01065	0.9999	5–200	2	5	87.5	13.9	80.6	3.0	93.6	1.2	LC
151	Mevinphos	C7H13O6P	3.62	225.05230	127.01547	0.9992	2–200	1	2	70.3	11.8	118.8	10.5	90.0	7.6	LC
152	Mirex	C10Cl12	29.05	271.80963	273.80667	0.9999	1–200	1	1	64.6	2.1	62.6	6.0	57.1	2.9	GC
153	Monocrotophos	C7H14NO5P	2.77	224.06824	58.02874	0.9995	1–200	1	1	99.3	17.3	110.1	12.7	81.7	3.4	LC
154	Myclobutanil	C15H17ClN4	10.56	289.12145	70.03997	0.9996	5–200	1	5	110.7	15.3	93.8	6.3	83.1	5.2	LC
155	Napropamide	C17H21NO2	11.63	272.16451	171.08044	0.9999	1–200	1	1	84.6	2.7	94.7	2.1	83.9	1.6	LC
156	Norflurazon	C12H9ClF3N3O	7.06	304.04590	140.03062	0.9998	1–200	1	1	87.7	2.8	95.0	2.7	82.4	1.0	LC
157	Omethoate	C5H12NO4PS	2.08	214.02974	182.98755	0.9988	1–200	1	1	85.5	8.5	92.2	5.6	80.2	2.7	LC
158	Oxadiazon	C15H18Cl2N2O3	25.39	174.95862	258.03214	0.9999	2–200	2	2	98.2	15.3	81.3	17.4	84.2	2.1	GC
159	Oxadixyl	C14H18N2O4	4.99	279.13393	132.08078	0.9999	1–200	1	1	84.8	19.7	117.3	11.0	93.9	3.3	LC
160	Paclobutrazol	C15H20ClN3O	25.79	236.05852	125.01525	0.9996	2–200	1	2	86.4	2.7	82.0	9.3	81.8	2.6	GC
161	Pentachloroaniline	C6H2Cl5N	18.83	264.85950	266.85657	0.9975	1–200	1	1	72.1	6.9	70.9	5.6	71.5	1.6	GC
162	Pentachloroanisole	C7H3Cl5O	14.82	264.83569	279.85919	0.9945	1–200	1	1	72.6	9.2	71.0	2.6	71.2	1.5	GC
163	Penthiopyrad	C16H20F3N3OS	14.47	360.13620	256.03506	0.9998	1–200	1	1	97.1	8.3	91.1	4.1	83.4	1.1	LC
164	Phenthoate	C12H17O4PS2	14.95	321.03786	79.05423	0.9999	2–200	2	2	84.7	11.0	74.6	18.2	100.6	7.8	LC
165	Phorate-Sulfone	C7H17O4PS3	8.56	293.00970	96.95076	0.9933	5–200	5	5	76.4	0.5	70.5	11.2	70.3	11.8	LC
166	Phorate-Sulfoxide	C7H17O3PS3	6.30	277.01502	96.95076	0.9998	1–200	1	1	99.7	7.2	97.6	5.5	85.6	1.2	LC
167	Phosalone	C12H15ClNO4PS2	15.96	367.99414	110.99960	0.9929	20–200	20	20	118.3	12.9	116.0	11.7	88.8	2.4	LC
168	Phosphamidon	C10H19ClNO5P	4.68	300.07621	127.01547	0.9997	1–200	1	1	88.2	7.1	91.8	3.1	84.7	1.2	LC
169	Phoxim	C12H15N2O3PS	15.98	299.06138	77.03889	0.9917	10–200	5	10	71.1	10.9	74.2	17.6	110.9	14.9	LC
170	Picoxystrobin	C18H16F3NO4	14.65	368.11042	145.06479	0.9993	1–200	1	1	101.0	16.6	94.6	8.8	84.0	5.2	LC
171	Piperonyl butoxide	C19H3O05	17.06	356.24230	119.08553	0.9994	1–200	1	1	93.5	10.4	82.4	7.7	78.2	1.6	LC

Table 1. *Cont.*

No	Compound	Formula	RT (Min)	Quantitative Ion	Qualitative Ion	R²	Linearity (ng/g)	SDL (ng/g)	LOQ (ng/g)	1-LOQ REC (%)	1-LOQ RSD (%)	2-LOQ REC (%)	2-LOQ RSD (%)	10-LOQ REC (%)	10-LOQ RSD (%)	Detecting Instrument
172	Pirimicarb	C11H18N4O2	4.41	239.15025	72.04439	0.9975	1–200	1	1	78.3	14.4	95.3	4.3	78.0	3.7	LC
173	Pirimiphos-methyl	C11H20N3O3PS	15.87	306.10358	67.02908	0.9999	1–200	1	1	86.0	4.2	91.7	1.1	80.5	0.8	LC
174	Pretilachlor	C17H26ClNO2	16.17	312.17248	252.11497	0.9999	1–200	1	1	116.7	4.8	88.7	9.4	82.8	3.9	LC
175	Prochloraz	C15H16Cl3N3O2	13.20	376.03809	70.02874	0.9998	1–200	1	1	82.4	6.7	97.2	7.6	79.0	2.6	LC
176	Profenofos	C11H15BrClO3PS	16.14	372.94242	96.95094	0.9984	2–200	2	2	94.5	15.6	92.5	6.5	98.6	3.3	LC
177	Prometryn	C10H19N5S	8.73	242.14339	68.02432	0.9997	1–200	1	1	82.1	2.7	89.7	1.9	79.7	2.0	LC
178	Propamocarb	C9H20N2O2	2.18	189.15975	74.02366	0.9983	1–200	1	1	69.3	17.2	90.3	13.7	79.9	8.3	LC
179	Propanil	C9H9Cl2NO	7.97	218.01340	127.01784	0.9996	5–200	2	5	71.1	6.8	70.4	7.0	88.1	1.7	LC
180	Propaphos	C13H21O4PS	13.10	305.09709	44.97935	0.9998	1–200	1	1	83.3	5.4	83.7	2.8	81.4	1.5	LC
181	Propargite	C19H26O4S	18.28	368.18860	57.06988	0.9910	5–200	5	5	84.8	15.8	96.3	12.5	116.2	19.9	LC
182	Propazine	C9H16ClN5	8.11	230.11670	146.02280	0.9992	1–200	1	1	82.0	1.8	99.6	4.1	80.8	2.7	LC
183	Propiconazole	C15H17Cl2N3O2	13.23	342.07706	69.06988	0.9999	5–200	1	1	85.5	5.2	87.0	5.8	77.6	3.4	LC
184	Propyzamide	C12H11Cl2NO	11.01	256.02905	189.98210	0.9989	1–200	1	5	82.8	4.2	82.3	10.9	92.2	4.1	LC
185	Prothioconazole	C14H15Cl2N3OS	12.48	344.03860	102.01205	0.9942	5–200	5	5	70.2	0.8	117.9	18.9	70.4	10.6	LC
186	Prothioconazole-desthio	C14H15Cl2N3O	10.35	312.06640	70.03997	0.9999	1–200	1	1	87.0	6.6	89.6	5.0	80.2	1.7	LC
187	Pymetrozine	C10H11N5O	2.04	218.10364	105.04472	0.9943	1–200	1	1	113.0	9.8	81.7	8.7	71.3	11.5	LC
188	Pyraclostrobin	C19H18ClN3O4	15.40	388.10586	194.08118	0.9999	0.2–200	0.2	0.2	105.8	18.2	104.8	13.2	84.5	2.4	LC
189	Pyridaben	C19H25ClN2OS	18.83	365.14489	147.11682	0.9988	1–200	1	1	114.4	19.2	80.5	9.1	70.2	2.5	LC
190	Pyridaphenthion	C14H17N2O4PS	11.59	341.07194	92.04979	0.9998	1–200	1	1	72.5	9.5	92.7	6.8	85.6	1.5	LC
191	Pyrimethanil	C12H13N3	7.56	200.11822	77.03857	0.9995	5–200	1	5	84.9	6.6	76.3	2.2	90.8	3.4	LC
192	Pyriproxyfen	C20H19NO3	17.50	322.14377	96.04439	0.9998	1–200	1	1	87.7	15.8	86.4	3.4	76.5	3.5	LC
193	Quinalphos	C12H15N2O3PS	14.00	299.06138	96.95076	0.9999	1–200	1	1	90.3	10.2	100.0	2.9	79.9	2.4	LC
194	Quinoxyfen	C15H8Cl2FNO	16.79	308.00397	196.97887	0.9998	1–200	1	1	72.8	1.8	79.1	3.8	70.2	4.3	LC
195	Quintozene	C6Cl5NO2	16.21	236.84077	294.83371	0.9972	1–200	1	1	82.3	18.0	73.6	10.0	82.5	16.0	GC
196	Quizalofop-ethyl	C19H17ClN2O4	16.62	373.09496	91.05423	0.9997	1–200	1	1	105.5	15.8	103.8	9.9	76.0	0.9	LC
197	Saflufenacil	C17H17ClF4N4O5S	10.90	501.06170	348.99976	0.9994	1–200	1	1	112.6	17.7	83.8	17.3	82.7	6.0	LC
198	Simazine	C7H12ClN5	5.00	202.08540	68.02432	0.9997	1–200	1	1	93.1	1.7	96.5	5.2	84.2	2.3	LC
199	Spinosyn D	C42H67NO10	15.36	746.48377	142.12263	0.9998	1–200	1	1	88.0	8.6	98.9	9.1	80.1	5.2	LC
200	Spirodiclofen	C21H24Cl2O4	18.97	411.11244	71.08553	0.9992	0.5–200	0.5	0.5	73.9	19.9	119.8	12.3	103.1	13.8	LC
201	Spirotetramat	C21H27NO5	10.10	374.19620	302.17508	0.9996	5–200	2	5	81.0	12.2	70.5	7.6	77.3	5.3	LC
202	Spirotetramat-enol	C18H23NO3	5.29	302.17580	216.10190	0.9996	1–200	1	1	84.6	7.0	83.2	2.8	75.5	3.0	LC
203	Spirotetramat-enol-glucoside	C24H33NO8	2.86	464.22790	216.10190	0.9926	1–200	1	1	120.8	8.5	134.3	2.0	185.9	11.2	LC
204	Spiroxamine	C18H35NO2	8.76	298.27406	100.11208	0.9990	1–200	1	1	91.3	12.4	80.9	3.1	78.2	3.5	LC
205	Sulfentrazone	C11H10Cl2F2N4O3S	6.34	386.98915	306.99435	0.9988	5–200	5	5	76.7	5.3	83.6	4.9	94.0	4.3	LC
206	Sulfotep	C8H20O5P2S2	15.67	322.02219	237.92828	1.0000	1–200	1	1	95.4	16.1	91.2	9.3	79.1	9.5	GC

Table 1. Cont.

No	Compound	Formula	RT (Min)	Quantitative Ion	Qualitative Ion	R²	Linearity (ng/g)	SDL (ng/g)	LOQ (ng/g)	1-LOQ REC (%)	1-LOQ RSD (%)	2-LOQ REC (%)	2-LOQ RSD (%)	10-LOQ REC (%)	10-LOQ RSD (%)	Detecting Instrument
207	Sulfoxaflor	C10H10F3N3OS	4.48	278.05690	154.04628	0.9983	2–200	2	2	91.0	8.9	72.7	5.3	100.7	5.8	LC
208	Sulprofos	C12H19O2PS3	17.99	323.03575	218.96979	0.9997	2–200	2	2	72.4	14.0	89.1	10.9	86.6	5.5	LC
209	Tebuconazole	C16H22ClN3O	11.75	308.15240	70.03997	0.9999	2–200	1	2	95.8	5.9	80.8	6.7	87.3	6.0	LC
210	Tebufenozide	C22H28N2O2	13.92	353.22235	133.06479	0.9984	2–200	1	2	113.1	9.4	58.5	18.1	85.0	17.5	LC
211	Terbufos	C9H21O2PS3	17.05	289.05141	57.06988	0.9981	10–200	2	10	88.5	16.1	90.9	7.5	85.0	15.4	LC
212	Terbufos-Sulfone	C9H21O4PS3	11.57	321.04120	275.05353	0.9982	2–200	2	2	114.0	16.7	92.9	9.6	98.1	3.4	LC
213	Terbufos-Sulfoxide	C9H21O3PS3	8.23	305.04650	130.93848	0.9999	1–200	1	1	114.9	13.2	106.3	12.6	79.7	5.2	LC
214	Terbumeton	C10H19N5O	17.40	210.13493	169.09581	0.9986	1–200	1	1	111.8	18.6	79.2	6.9	85.4	2.4	LC
215	Terbuthylazine	C9H16ClN5	8.82	230.11670	174.05409	0.9998	1–200	1	1	89.3	18.9	98.2	8.4	79.8	3.4	GC
216	Terbutryn	C10H19N5S	9.10	242.14379	186.08080	0.9992	1–200	1	1	82.3	3.6	90.0	2.3	76.2	4.4	LC
217	Tetramethrin	C19H25NO4	17.04	332.18560	164.07060	0.9988	5–200	2	5	97.8	13.0	92.4	4.4	98.6	5.5	LC
218	Thiabendazole	C10H7N3S	2.90	202.04334	131.06038	0.9999	1–200	1	1	74.3	3.9	82.4	5.1	73.1	1.8	LC
219	Thiacloprid	C10H9ClN4S	4.51	253.03092	126.00867	0.9993	1–200	1	1	87.3	5.1	98.5	1.7	82.0	2.0	LC
220	Thiamethoxam	C8H10ClN5O3S	3.13	292.02656	131.96643	0.9950	1–200	1	1	79.4	11.6	81.3	5.6	74.7	6.9	LC
221	Thiobencarb	C12H16ClNOS	15.15	258.07150	125.01525	0.9985	2–200	2	2	84.6	18.6	89.5	13.2	90.6	1.3	LC
222	Thiophanate-methyl	C12H14N4O4S2	5.43	343.05292	151.03244	0.9992	1–200	1	1	102.4	5.6	82.9	2.0	73.9	3.0	LC
223	Tolfenpyrad	C21H22ClN3O2	16.91	384.14770	197.09608	0.9998	1–200	1	1	82.0	13.2	79.0	3.6	76.3	8.1	LC
224	Trans-Chlordane	C10H6Cl8	23.38	372.82544	374.82251	0.9997	1–200	1	1	78.6	6.2	71.2	6.4	70.6	4.5	GC
225	Triadimefon	C14H16ClN3O2	11.17	294.10038	57.06988	0.9996	1–200	1	1	90.9	10.0	85.9	8.2	83.1	3.9	LC
226	Triadimenol	C14H18ClN3O2	8.54	296.11580	70.03997	0.9998	5–200	5	5	90.6	14.0	74.5	11.0	84.3	4.3	LC
227	Triazophos	C12H16N3O3PS	12.72	314.07228	119.06037	0.9980	1–200	1	1	84.1	2.2	61.2	2.3	83.3	1.9	LC
228	Trichlorfon	C4H8Cl3O4P	3.33	256.92985	78.99452	0.9992	10–200	5	10	77.3	13.4	91.6	8.5	84.2	7.1	LC
229	Trifloxystrobin	C20H19F3N2O4	16.67	409.13697	145.02596	0.9992	1–200	1	1	88.2	3.8	89.0	3.7	81.4	1.3	LC
230	Triflumizole	C15H15ClF3N3O	14.98	346.09290	69.04472	1.0000	1–200	1	1	88.5	6.5	90.9	4.6	80.5	2.0	LC
231	Trifluralin	C13H16F3N3O4	15.26	264.02267	306.06961	0.9981	2–200	2	2	79.1	7.0	71.3	11.1	80.3	6.4	GC
232	Trinexapac-ethyl	C13H16O5	7.54	253.10705	69.03349	0.9996	5–200	5	5	75.4	9.5	60.1	12.4	73.4	3.4	LC
233	Uniconazole	C15H18ClN3O	10.58	292.12130	70.03997	0.9999	1–200	1	1	60.9	16.4	84.6	8.3	79.7	1.7	LC
234	Vinclozolin	C12H9Cl2NO3	20.63	212.00284	186.95862	0.9962	2–200	2	2	67.7	12.7	80.4	4.4	83.7	4.9	LC
235	Warfarin	C19H16O4	8.91	309.11214	163.03897	0.9997	0.5–200	0.5	0.5	98.7	15.0	99.5	6.0	91.0	6.7	LC
236	Zoxamide	C14H16Cl3NO2	14.92	336.03194	186.97119	0.9997	1–200	1	1	92.8	7.1	96.0	4.2	83.0	2.0	LC
237	Endrin	C12H8Cl6O	25.22	316.90341	262.85641	0.9962	5–200	5	5	63.7	19.8	76.5	5.5	78.2	4.5	GC

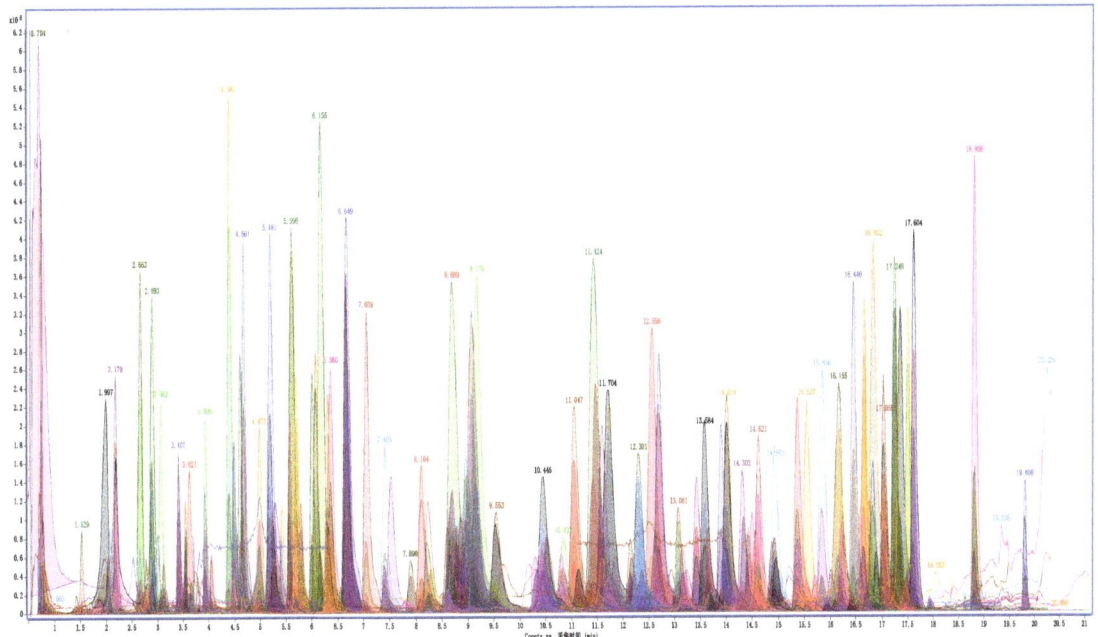

Figure 7. Overlay extraction ion chromatograms of LC-Q TOF/MS of cottonseed hull sample at spiking level of 200 μg/kg.

The calibration curve was plotted using the matrix matching calibration method and the target analytes at 10 spiked levels (0.2, 0.5, 1, 2, 5, 10, 20, 50, 100, and 200 μg/kg) were spiked to the blank cottonseed hull sample. The linear ranges of 237 pesticide analytes were 1–200 μg/L. All target pesticides showed good linearity in the concentration range, and R^2 was greater than 0.99, indicating that this method could meet the requirements of quantitative analysis.

3.6.2. Recovery and Precision

The recovery and precision of the method was evaluated by spiked standard solutions at the levels of 1-, 2-, and 10-times LOQ for the cottonseed hull samples with six parallels at each spiked level. The results are shown in Figure 8. At the levels of 1-, 2-, and 10-times LOQ, the recoveries of the 237 pesticides in the range of 70–120% were 91.6%, 92.8%, and 94.5%, respectively, and the RSD of all the pesticides was less than 20%, indicating that the method had satisfactory recovery and precision.

Among the 237 pesticides, 60 pesticides were detected by two detection techniques, and most of them showed similar performance; however, individual pesticides were different in the two techniques. For example, the average recovery (81.2%) of clodinafop-propargyl detected by GC-QTOF/MS was lower than that (95.7%) detected by LC-QTOF/MS. In terms of precision, the RSD (10.8%) of the compound detected by GC-QTOF/MS was higher than that (4.8%) detected by LC-QTOF/MS. For Propiconazole, the average recovery and RSD of GC-QTOF/MS (89.0%, 5.5%) were better than those of LC-QTOF/MS (80.0%, 6.4%). Therefore, appropriate detection techniques should be selected in pesticide residue analysis, especially when compounds are suitable for these two detection techniques.

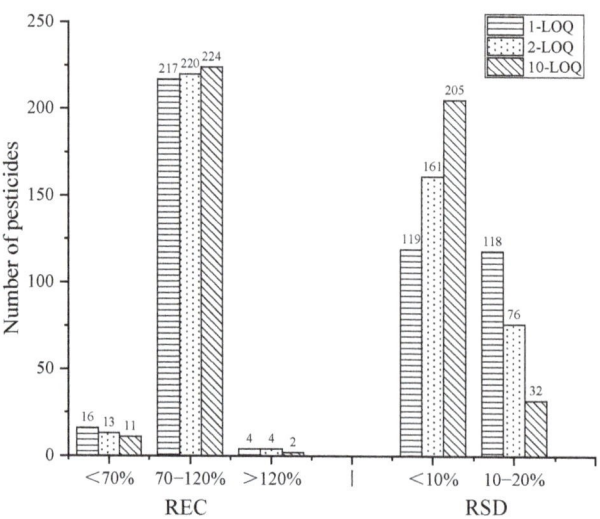

Figure 8. The recovery and RSD of the target pesticides at three spiked levels.

3.7. Analysis of Real Samples

The established method was applied to the analysis of 11 real cottonseed hull samples collected from several domestic pastures. The results showed that three pesticide residues were found in 11 cottonseed hull samples (butylate (three times), fenbuconazole (three times), and Diuron (two times)), with concentrations ranging from 10 to 28 µg/kg and above the LOQ. The determined three pesticides were slightly hazardous, according to WHO [32]. This method can be used for high-throughput trace detection of pesticide residues in cottonseed hull samples and improve the ability of risk-screening.

4. Conclusions

In this work, GC-QTOF/MS and LC-QTOF/MS were used to develop a high throughput method for qualitative screening and quantitative analysis of 237 pesticides in the cottonseed hull matrix. The modified QuEChERS extraction process seems to effectively eliminate the interference caused by the oily matrix, and the SDL, LOQ, recovery, and precision of the analysis method were verified under optimal conditions. In addition, compared with other methods for the oily matrix, this method has the advantages of being fast and simple, with high throughput and low solvent consumption. The results showed that the developed method could be applied to the screening of pesticide residues in the cottonseed hull matrix, effectively and generally.

Author Contributions: Conceptualization, H.C. and C.F.; methodology, H.C.; validation, K.T., Y.X. and X.W.; investigation, S.H. and Y.L.; resources, K.T.; data curation, Y.X.; writing—original draft preparation, K.T.; writing—review and editing, H.C., X.W. and C.F.; supervision, M.L. and W.W.; project administration, H.C. All authors have read and agreed to the published version of the manuscript.

Funding: This work was financially supported by the Science and Technology Project of the State Administration for Market Regulation (2021MK165).

Institutional Review Board Statement: Not applicable.

Informed Consent Statement: Not applicable.

Data Availability Statement: Not applicable.

Conflicts of Interest: The authors declare no conflict of interest.

References

1. Moretti, D.B.; Jimenez, C.R.; Trinca, H.M.; Machado-Neto, R.; Louvandini, H. Cottonseed feeding changes oxidative stress markers in ewes during the peripartum period and increases the quality of colostrum. *Vet. J.* **2019**, *247*, 32–37. [CrossRef] [PubMed]
2. Eiras, C.E.; Guerrero, A.; Valero, M.V.; Pardo, J.A.; Ornaghi, M.G.; Rivaroli, D.C.; Sanudo, C.; Prado, I.N. Effects of cottonseed hull levels in the diet and ageing time on visual and sensory meat acceptability from young bulls finished in feedlot. *Animal* **2017**, *11*, 529–537. [CrossRef] [PubMed]
3. Machado, P.A.S.; Valadares, S.D.; Valadares, R.F.D.; Paulino, M.F.; Pina, D.D.; Paixao, M.L. Nutritional and productive parameters of beef cattle on pasture fed different amounts of supplement Parâmetros nutricionais e produtivos em bovinos de corte a pasto alimentados com diferentes quantidades de suplemento. *Rev. Bras. Zootec.* **2011**, *40*, 1303–1312. [CrossRef]
4. Buah-Kwofie, A.; Humphries, M.S. Validation of a modified QuEChERS method for the analysis of organochlorine pesticides in fatty biological tissues using two-dimensional gas chromatography. *J. Chromatogr. B* **2018**, *1105*, 85–92. [CrossRef]
5. Shi, Z.H.; Zhang, S.L.; Huai, Q.R.; Xu, D.; Zhang, H.Y. Methylamine-modified graphene-based solid phase extraction combined with UPLC-MS/MS for the analysis of neonicotinoid insecticides in sunflower seeds. *Talanta* **2017**, *162*, 300–308. [CrossRef]
6. Walorczyk, S.; Drozdzynski, D. Improvement and extension to new analytes of a multi-residue method for the determination of pesticides in cereals and dry animal feed using gas chromatography–tandem quadrupole mass spectrometry revisited. *J. Chromatogr. A* **2012**, *1251*, 219–231. [CrossRef]
7. David, F.; Devos, C.; Dumont, E.; Yang, Z.; Sandra, P.; Huertas-Perez, J.F. Determination of pesticides in fatty matrices using gel permeation clean-up followed by GC-MS/MS and LC-MS/MS analysis: A comparison of low- and high-pressure gel permeation columns. *Talanta* **2017**, *165*, 201–210. [CrossRef]
8. Xue, J.Y.; Li, H.C.; Liu, F.M.; Jiang, W.Q.; Chen, X.C. Determination of strobilurin fungicides in cotton seed by combination of acetonitrile extraction and dispersive liquidliquid microextraction coupled with gas chromatography. *J. Sep. Sci.* **2014**, *37*, 845–852. [CrossRef]
9. Zhan, J.; Li, J.D.; Liu, D.H.; Liu, C.; Yang, G.G.; Zhou, Z.Q.; Wang, P. A simple method for the determination of organochlorine pollutants and the enantiomers in oil seeds based on matrix solid-phase dispersion. *Food Chem.* **2016**, *194*, 319–324. [CrossRef]
10. Piao, H.L.; Jiang, Y.X.; Li, X.P.; Ma, P.Y.; Wang, X.H.; Song, D.Q.; Sun, Y. Matrix solid-phase dispersion coupled with hollow fiber liquid phase microextraction for determination of triazine herbicides in peanuts. *J. Sep. Sci.* **2019**, *42*, 2123–2130. [CrossRef]
11. Jiang, Y.P.; Li, Y.J.; Jiang, Y.T.; Li, J.G.; Pan, C.P. Determination of multiresidues in rapeseed, rapeseed oil, and rapeseed meal by acetonitrile extraction, low-temperature cleanup, and detection by liquid chromatography with tandem mass spectrometry. *J. Agric. Food Chem.* **2012**, *60*, 5089–5098. [CrossRef] [PubMed]
12. Theurillat, X.; Dubois, M.; Huertas-Pérez, J.F. A multi-residue pesticide determination in fatty food commodities by modified QuEChERS approach and gas chromatography-tandem mass spectrometry. *Food Chem.* **2021**, *353*, 129039. [CrossRef] [PubMed]
13. Guan, W.B.; Li, Z.N.; Zhang, H.Y.; Hong, H.J.; Rebeyev, N.; Ye, Y.; Ma, Y.Q. Amine modified graphene as reversed-dispersive solid phase extraction materials combined with liquid chromatography–tandem mass spectrometry for pesticide multi-residue analysis in oil crops. *J. Chromatogr. A* **2013**, *1286*, 1–8. [CrossRef] [PubMed]
14. Rutkowska, E.; Ozowicka, B.; Kaczyński, P. Compensation of matrix effects in seed matrices followed by gas chromatography-tandem mass spectrometry analysis of pesticide residues. *J. Chromatogr. A* **2020**, *1614*, 460738. [CrossRef]
15. Koesukwiwat, U.; Lehotay, S.J.; Mastovska, K.; Dorweiler, K.J.; Leeipatpiboon, N. Extension of the QuEChERS Method for Pesticide Residues in Cereals to Flaxseeds, Peanuts, and Doughs. *J. Agric. Food Chem.* **2010**, *58*, 5950–5958. [CrossRef]
16. Shinde, R.; Pardeshi, A.; Dhanshetty, M.; Anastassiades, M.; Banerjee, K. Development and validation of an analytical method for the multiresidue analysis of pesticides in sesame seeds using liquid- and gas chromatography with tandem mass spectrometry. *J. Chromatogr. A* **2021**, *1652*, 462346. [CrossRef]
17. Gonzalez-Curbelo, M.A.; Socas-Rodriguez, B.; Herrera-Herrera, A.V.; Gonzalez-Salamo, J.; Hernandez-Borges, J.; Rodriguez-Delgado, M.A. Evolution and applications of the QuEChERS method. *TrAC Trends Anal. Chem.* **2015**, *71*, 169–185. [CrossRef]
18. Farre, M.; Kantiani, L.; Petrovic, M.; Perez, S.; Barcelo, D. Achievements and future trends in the analysis of emerging organic contaminants in environmental samples by mass spectrometry and bioanalytical techniques. *J. Chromatogr. A* **2012**, *1259*, 86–99. [CrossRef]
19. Dankyi, E.; Carboo, D.; Gordon, C.; Fomsgaard, I.S. Application of the QuEChERS Procedure and LC-MS/MS for the Assessment of Neonicotinoid Insecticide Residues in Cocoa Beans and Shells. *J. Food Compos. Anal.* **2016**, *44*, 149–157. [CrossRef]
20. Chawla, S.; Patel, H.K.; Vaghela, K.M.; Pathan, F.K.; Gor, H.N.; Patel, A.R.; Shah, P.G. Development and validation of multiresidue analytical method in cotton and groundnut oil for 87 pesticides using low temperature and dispersive cleanup on gas chromatography and liquid chromatography-tandem mass spectrometry. *Anal. Bioanal. Chem.* **2016**, *408*, 983–997. [CrossRef]
21. Naik, R.H.; Pallavi, M.S.; Bheemanna, M.; PavanKumar, K.; Reddy, V.C.S.; Nidoni, R.U.; Paramasivam, M.; Yadav, S. Simultaneous determination of 79 pesticides in pigeonpea grains using GC-MS/MS and LC-MS/MS. *Food Chem.* **2021**, *347*, 128986. [CrossRef]
22. Elbashir, A.A.; Aboul-Enein, H.Y. Application of gas and liquid chromatography coupled to time-of-flight mass spectrometry in pesticides: Multiresidue analysis. *Biomed. Chromatogr.* **2018**, *32*, e4038. [CrossRef] [PubMed]
23. Wang, J.; Chow, W.; Leung, D. Applications of LC/ESI-MS/MS and UHPLC QqTOF MS for the determination of 148 pesticides in fruits and vegetables. *J. AOAC Int.* **2011**, *396*, 1513–1538. [CrossRef] [PubMed]

24. Lozano, A.; Rajski, L.; Ucles, S. Evaluation of zirconium dioxide-based sorbents to decrease the matrix effect in avocado and almond multiresidue pesticide analysis followed by gas chromatography tandem mass spectrometry. *Talanta* **2014**, *118*, 68–83. [CrossRef]
25. Pang, G.F.; Chang, Q.Y.; Bai, R.B.; Fan, C.L.; Zhang, Z.J.; Yan, H.Y.; Wu, X.Q. Simultaneous Screening of 733 Pesticide Residues in Fruits and Vegetables by a GC/LC-Q-TOFMS Combination Technique. *Engineering* **2020**, *6*, 432–441. [CrossRef]
26. Dos-Reis, R.R.; Sampaio, S.C.; De Melo, E.B. The effect of different log P algorithms on the modeling of the soil sorption coefficient of nonionic pesticides. *Water Res.* **2013**, *47*, 5751–5759. [CrossRef]
27. Mol, H.G.J.; Plaza-BolanOs, P.; Zomer, P.; De Rijk, T.C.; Stolker, A.A.M.; Mulder, P.P.J. Toward a generic extraction method for simultaneous determination of pesticides, mycotoxins, plant toxins, and veterinary drugs in feed and food matrixes. *Anal. Chem.* **2008**, *80*, 9450–9459. [CrossRef]
28. Tuzimski, T.; Rejczak, T. Application of HPLC-DAD after SPE/QuEChERS with ZrO_2-based sorbent in d-SPE clean-up step for pesticide analysis in edible oils. *Food Chem.* **2016**, *190*, 71–79. [CrossRef]
29. Lagunas-Allue, L.; Sanz-Asensio, J.; Martínez-Soria, M.T. Comparison of four extraction methods for the determination of fungicide residues in grapes through gas chromatography-mass spectrometry. *J. Chromatogr. A* **2012**, *1270*, 62–71. [CrossRef]
30. Ucles, S.; Lozano, A.; Sosa, A.; Vazquez, P.P.; Valverde, A.; Fernandez-Alba, A.R. Matrix interference evaluation employing GC and LC coupled to triple quadrupole tandem mass spectrometry. *Talanta* **2017**, *174*, 72–81. [CrossRef]
31. Kuzukiran, O.; Simsek, I.; Yorulmaz, T.; Yurdakok-Dikmen, B.; Ozkan, O.; Filazi, A. Multiresidues of environmental contaminants in bats from Turkey. *Chemosphere* **2021**, *282*, 131022. [CrossRef] [PubMed]
32. World Health Organization. *The WHO Recommended Classification of Pesticides by Hazard and Guidelines to Classification 2019*; World Health Organization: Geneva, Switzerland, 2020.

 separations MDPI

Article

Development of a High-Throughput Screening Analysis for 195 Pesticides in Raw Milk by Modified QuEChERS Sample Preparation and Liquid Chromatography Quadrupole Time-of-Flight Mass Spectrometry

Xingqiang Wu [1], Kaixuan Tong [1], Changyou Yu [2], Shuang Hou [2], Yujie Xie [1], Chunlin Fan [1], Hui Chen [1,*], Meiling Lu [3] and Wenwen Wang [3]

[1] Key Laboratory of Food Quality and Safety for State Market Regulation, Chinese Academy of Inspection & Quarantine, No. 11, Ronghua South Road, Beijing 100176, China; xingqiangheda@163.com (X.W.); tongkx@caiq.org.cn (K.T.); xieyj@caiq.org.cn (Y.X.); caiqfcl@163.com (C.F.)

[2] Laboratory of Heilongjiang Feihe Dairy Co., Ltd., Qiqihar 164800, China; yuchangyou@feihe.com (C.Y.); houshuang@feihe.com (S.H.)

[3] Agilent Technologies (China) Limited, Beijing 100102, China; mei-ling.lu@agilent.com (M.L.); wen-wen_wang@agilent.com (W.W.)

* Correspondence: chenh@caiq.org.cn

Abstract: This study aimed to develop a simple, high-throughput method based on modified QuEChERS (quick, easy, cheap, effective, rugged, and safe) followed by liquid chromatography quadrupole time-of-flight mass spectrometry (LC-Q-TOF/MS) for the rapid determination of multi-class pesticide residues in raw milk. With acidified acetonitrile as the extraction solvent, the raw milk samples were pretreated with the modified QuEChERS method, including extraction, salting-out, freezing, and clean-up processes. The target pesticides were acquired in a positive ion electrospray ionization mode and an All ions MS/MS mode. The developed method was validated, and good performing characteristics were achieved. The screening detection limits (SDL) and limits of quantitation (LOQ) for all the pesticides ranged within 0.1–20 and 0.1–50 μg/kg, respectively. The recoveries of all analytes ranged from 70.0% to 120.0% at three spiked levels (1 × LOQ, 2 × LOQ, and 10 × LOQ), with relative standard deviations less than 20.0%. The coefficient of determination was greater than 0.99 within the calibration linearity range for the detected 195 pesticides. The method proved the simple, rapid, high throughput screening and quantitative analysis of pesticide residues in raw milk.

Keywords: raw milk 1; pesticides 2; screening 3; QuEChERS 4; high-throughput 5

Citation: Wu, X.; Tong, K.; Yu, C.; Hou, S.; Xie, Y.; Fan, C.; Chen, H.; Lu, M.; Wang, W. Development of a High-Throughput Screening Analysis for 195 Pesticides in Raw Milk by Modified QuEChERS Sample Preparation and Liquid Chromatography Quadrupole Time-of-Flight Mass Spectrometry. *Separations* 2022, 9, 98. https://doi.org/10.3390/separations9040098

Academic Editor: Chiara Emilia Cordero

Received: 28 March 2022
Accepted: 11 April 2022
Published: 12 April 2022

Publisher's Note: MDPI stays neutral with regard to jurisdictional claims in published maps and institutional affiliations.

Copyright: © 2022 by the authors. Licensee MDPI, Basel, Switzerland. This article is an open access article distributed under the terms and conditions of the Creative Commons Attribution (CC BY) license (https://creativecommons.org/licenses/by/4.0/).

1. Introduction

Milk is considered an important part of a healthy diet, providing essential nutrients and energy. High-quality raw milk is required by dairy factories to make dairy products, such as cheese, yogurt, and cream [1]. Once the raw milk is defective, it cannot be improved in the subsequent processing, which may have far-reaching effects. Currently, China is one of the world's largest producing and consuming countries of milk and dairy products, with the per capita consumption of milk in China increasing from 4.89 kg in 1997 to 19.2 kg in 2019 [2]. The quality and safety of milk and its products are of a great concern to both the government and consumers [3]. Meanwhile, the contamination of milk with pesticide residues is a severe concern in many countries [4–6]. Pesticide residues in milk may come from direct or indirect sources such as feeding animals from contaminated forage grass, feeding and drinking water, and various pesticides used to treat pests, pathogens, and fungal diseases [7]. Through the above pathways, these pesticide residues inevitably accumulate in animals. They are transferred to secreted milk, with serious health hazards likely to occur as humans consume contaminated milk or dairy products [8,9]. Hence, it is necessary to ascertain pesticide residues in milk to ensure safe dietary intake.

To ensure food safety, several organizations and countries, such as the European Commission [10] and China [11], have established maximum residue limits (MRL) for various pesticides in milk. Therefore, to meet these requirements, there is an increasing need for an effective analytical method for simultaneous qualitative and quantitative screening of pesticide residues in milk. The current reported methods for the analysis of multi-residue pesticides in milk use different detection techniques, such as high-performance liquid chromatography with diode-array detection (HPLC-DAD) [12], gas chromatography–electron capture detection (GC-ECD) [13], gas chromatography–mass spectrometry (GC-MS) [14], gas chromatography–tandem mass spectrometry (GC-MS/MS) [15,16], and liquid chromatography–tandem mass spectrometry (LC-MS/MS) [17–19]. Recently, liquid chromatography coupled with high-resolution mass spectrometry techniques (LC-HRMS) had been applied to determine pesticide residues in milk matrices [20,21]. LC-HRMS offered the ability to collect full scan spectra and accurate masses while acquiring and reprocessing data without prior compound-specific adjustments, enabling retrospective data analysis [22]. Hence, LC-HRMS has a strong competitive advantage compared with low-resolution mass spectrometry in the multi-residue analysis of compounds and has demonstrated great potential for non-targeted detection.

Although LC-HRMS demonstrates high sensitivity and accuracy in developing analytical methods, selecting a suitable sample preparation method is an important prerequisite for achieving multi-residue analysis. Milk is a complex matrix in which interfering components (e.g., proteins, fatty acids, and pigments) may play a role in suppressing the signal of pesticide residues. Therefore, effectively reducing matrix interference is crucial for determining pesticide residues in milk [23]. Different sample preparation methods for extracting pesticides from milk have been explored. These methods mainly include liquid–liquid extraction (LLE) [19,24], gel permeation chromatography (GPC) [15], solid-phase extraction (SPE) [5,25], dispersive solid-phase extraction (d-SPE) [21], and the QuEChERS (quick, easy, cheap, effective, rugged, and safe) method [13,14,16,18]. Among them, GPC and SPE are tedious and time-consuming to operate, which do not facilitate the processing of a large number of samples. Meanwhile, LE and d-SPE methods have a large background interference of the sample matrix after pretreatment, which causes a decrease in detection sensitivity of the analytical instrument [26]. QuEChERS is fast, safe, and low-cost in the aforementioned techniques, including extraction and purification steps. Compared to other sample preparation techniques, QuEChERS is simple to use and has efficiency improvement with good reproducibility and stability. The QuEChERS method has been widely used for the high-throughput analysis of chemical contaminants in various food products [27].

This work aimed to establish a simple and efficient pretreatment method for the simultaneous detection of multi-pesticide residues in raw milk using an advanced LC-Q-TOF/MS technique. The pretreatment procedure was optimized, including different extraction salts, purification sorbents, and freezing times. Meanwhile, this method's linearity, sensitivity, accuracy, precision, and matrix effect were fully evaluated. Finally, a simple and effective sample preparation procedure was established to determine 195 pesticide residues in raw milk combined with LC-Q-TOF/MS. Moreover, the validated method was employed to screen pesticide residues in actual raw milk samples from dairy farms.

2. Materials and Methods

2.1. Instrumentation

The liquid chromatography quadrupole time-of-flight mass spectrometry (1290–6550) was from Agilent Technologies (Santa Clara, CA, USA). Chromatographic separation was achieved on a chromatographic condition: equipped with a reversed-phase chromatography column (ZORBAX SB-C18 column 2.1 mm × 100 mm, 3.5 μm; Agilent Technologies, Santa Clara, CA, USA); mobile phase A is 5 mM ammonium acetate-0.1% formic acid-water;

mobile phase B is acetonitrile; gradient elution program, 0 min: 1%B, 3 min: 30%B, 6 min: 40%B, 9 min: 40%B, 15 min: 60%B, 19 min: 90%B, 23 min: 90%B, 23.01 min: 1%B, run after 4 min. The flow rate was set at 0.4 mL/min. The column temperature was 40 °C. The injection volume was 5 µL.

An Agilent Dual Jet Stream electrospray source was used on the Q-TOF in positive ionization mode. The conditions for mass spectrometry were set as follows: Scan mode: All ions MS/MS; capillary voltage was 4 kV; nebulizer gas was 0.14 MPa; drying gas temperature was at 325 °C with a flow rate of 12.0 L/min; sheath gas temperature was set at 375 °C with a flow rate of 11.0 L/min; Fragmentation voltage at 145 v. All Ions MS/MS mode parameter settings: acquisition range was m/z 50–1000, data acquisition rate is four spectra/s; collision energy was 0 eV at 0 min, and collision energy was set to 0, 15, and 35 eV in consecutive order after 0.5 min.

The mass spectrum information of 195 pesticide databases is shown in Table 1. PL602-L electronic balance was purchased from Mettler-Toledo Co., Ltd. (Zurich, Switzerland); N-112 Nitrogen evaporator concentrator was obtained from Organomation Associates (EVAP 112, Worcester, MA, USA); SR-2DS oscillator was obtained from Taitec company (Saitama, Japan); KDC-40 Low-speed centrifuge was obtained from Zonkia Group Corp., Ltd. (Hefei, China); Milli-Q ultrapure water machine was obtained from Millipore Co., Ltd. (Milford, MA, USA).

2.2. Reagents and Materials

Raw milk samples were collected from local dairy farms. All pesticide standards (purity grade, >98%) were obtained from Alta Company (Tianjin, China). Formic acid, ammonium acetate, acetonitrile, methanol (all LC-MS grade), and toluene (HPLC grade) were obtained from Fisher Scientific, Inc. (Fair Lawn, NJ, USA). Analytical grade forms of acetic acid, sodium chloride, anhydrous Na_2SO_4, trisodium citrate, disodium citrate, and anhydrous $MgSO_4$ were obtained from Shanghai Anpu Experimental Technology (Shanghai, China). The cleanup absorbents as octadecylsilane (C18) and primary secondary amine (PSA) were obtained from Tianjin Agela Technology (Tianjin, China).

2.3. Preparation of Standard Solutions

Standard stock solutions of individual pesticides were prepared in acetonitrile, methanol, or water to a concentration of 500–1000 mg/L. All stock solutions were stable for 6 months in a closed tea-colored volumetric flask at −20 °C. The 10 mg/L intermediate working solution and the working internal standard solution (Atrazine-D5) were prepared by diluting the stock solution with methanol. Working solutions were prepared daily by diluting a stock solution with all pesticides and used immediately after preparation.

2.4. Sample Preparation

The QuEChERS procedure entailed the following steps: 2.0 g of raw milk sample were weighed into the 50 mL tube. 16 mL of 1% acetic acid acetonitrile (v/v) was added, followed by EN salt (4 g $MgSO_4$, 1 g NaCl, 0.5 g disodium citrate, and 1 g trisodium citrate), vortexed for 1 min, and shaken for 2 min. After that, the sample tubes were frozen at −20 °C for 0.5 h and then centrifuged (4200 rpm) for 5 min. 5 mL of supernatant was again pipetted into a 15 mL clean-up tube (containing 500 mg $MgSO_4$ and 200 mg C18). The clean-up tube was vortexed for 5 s and then shaken for 2 min, followed by centrifugation at 4200 rpm for 5 min. Subsequently, 2 mL of the supernatant from the clean-up tube was pipetted into a 10 mL glass tube and evaporated to dryness in a 40 °C water bath with a gentle stream of nitrogen. Finally, 1 mL of acetonitrile/water (3:2, v/v) solution was used to redissolve the solution and pass it over the membrane for LC-Q-TOF/MS analysis.

Table 1. LC-Q-TOF/MS parameters and validation parameters for all target analytes in raw milk.

NO.	Compound	Formula	RT/Min	Quantitative Ion (m/z)	Production (m/z)	SDL (mg/kg)	LOQ (mg/kg)	MRL (mg/kg) European Union, China	R^2	1 × LOQ Rec. (%)	1 × LOQ RSD (%)	2 × LOQ Rec. (%)	2 × LOQ RSD (%)	10 × LOQ Rec. (%)	10 × LOQ RSD (%)
1	1-(2-chloro-4-(4-chlorophenoxy)phenyl)-2-(1H-1,2,4-triazol-1-yl)ethanol	$C_{16}H_{13}Cl_2N_3O_2$	10.16	350.0458	70.0400	20.0	20.0	—, —	0.9988	100.2	1.0	98.1	0.9	86.2	1.1
2	1-(2-Chloro-pyridin-5-yl-methyl)-2-imino-imidazolidine hydrochloride	$C_9H_{12}Cl_2N_4$	2.28	211.0745	90.0338	0.5	1.0	—, —	0.9990	94.4	18.7	82.2	6.7	101.0	16.8
3	1-methyl-3-(tetrahydro-3-furylmethyl) urea	$C_7H_{14}N_2O_2$	1.87	159.1128	58.0287	0.2	1.0	—, —	0.9926	96.2	7.6	98.7	14.3	104.2	11.7
4	3-(Trifluoromethyl)-1-methyl-1H-pyrazole-4-carboxamide	$C_6H_6F_3N_3O$	2.63	194.0536	134.0349	10.0	10.0	—, —	0.9932	94.8	14.7	106.4	8.5	85.8	6.6
5	5-hydroxy Imidacloprid	$C_9H_{10}ClN_5O_3$	3.05	272.0545	225.0538	2.0	5.0	—, —	0.9992	109.6	11.3	112.9	18.0	101.4	18.0
6	Acetamiprid	$C_{10}H_{11}ClN_4$	3.97	223.0745	126.0105	0.5	0.5	0.2, —	0.9994	77.9	5.8	84.6	10.4	103.9	7.5
7	Acetamiprid-N-desmethyl	$C_9H_9ClN_4$	3.62	209.0589	126.0105	0.2	1.0	—, —	0.9976	119.0	12.3	94.6	15.3	97.1	15.7
8	Acetochlor	$C_{14}H_{20}ClNO_2$	12.62	270.1255	133.0886	1.0	1.0	0.01, —	0.9989	83.4	18.5	119.8	6.8	101.7	10.5
9	Alachlor	$C_{14}H_{20}ClNO_2$	12.58	270.1255	238.0993	1.0	2.0	0.01, —	0.9989	118.6	7.4	98.4	3.2	94.7	2.2
10	Aldicarb-sulfone	$C_7H_{14}N_2O_4S$	2.66	223.0747	62.9899	10.0	20.0	—, —	0.9980	99.3	7.5	95.7	3.6	87.6	2.4
11	Allidochlor	$C_8H_{12}ClNO$	5.00	174.0680	98.0964	10.0	10.0	—, —	0.9968	71.4	16.8	85.9	6.5	72.1	17.6
12	Ametryn	$C_9H_{17}N_5S$	6.71	228.1277	68.0243	0.1	0.5	—, —	0.9973	96.2	2.8	98.0	1.7	100.2	1.4
13	Aminocyclopyrachlor	$C_8H_8ClN_3O_2$	0.76	214.0378	68.0495	10.0	10.0	—, —	0.9976	72.9	9.9	75.7	8.8	86.4	11.6
14	Aminopyralid	$C_6H_4Cl_2N_2O_2$	1.70	206.9723	160.9668	20.0	50.0	0.02, —	0.9973	70.0	8.9	76.0	6.7	83.0	5.3
15	Atrazine	$C_8H_{14}ClN_5$	6.44	216.1010	174.0541	0.1	0.1	—, —	0.9976	87.3	13.6	105.9	3.9	101.7	4.4
16	Avermectin	$C_{48}H_{72}O_{14}$	18.72	895.4814	751.4052	0.5	0.5	—, —	0.9993	87.4	7.4	108.6	3.8	92.7	4.7
17	Azoxystrobin	$C_{22}H_{17}N_3O_5$	11.17	404.1241	329.0795	0.1	0.1	0.01, —	0.9973	86.8	19.3	97.2	12.6	100.7	3.9
18	Benalaxyl	$C_{20}H_{23}NO_3$	14.11	326.1751	91.0542	0.2	0.5	0.02, —	0.9981	110.5	9.8	92.5	2.7	101.2	1.5
19	Benzovindiflupyr	$C_{18}H_{15}Cl_2F_2N_3O$	14.43	398.0640	159.0364	0.5	0.5	0.01, —	0.9985	93.6	7.0	107.6	4.2	100.7	2.0
20	Bioresmethrin	$C_{22}H_{26}O_3$	19.09	339.1955	143.0855	10.0	20.0	—, —	0.9905	103.3	17.6	80.5	11.7	82.1	9.8
21	Bitertanol	$C_{20}H_{23}N_3O_2$	12.77	338.1863	70.0400	10.0	10.0	0.01, —	0.9964	101.6	16.1	83.5	5.8	90.0	3.5
22	Boscalid	$C_{18}H_{12}Cl_2N_2O$	11.30	343.0399	271.0866	1.0	1.0	0.02, —	0.9989	116.4	8.3	105.6	8.9	104.1	13.6
23	Bromobutide	$C_{15}H_{22}BrNO$	13.80	312.0958	119.0855	1.0	2.0	—, —	0.9999	90.9	18.1	104.3	8.5	101.0	3.4
24	Bupirimate	$C_{13}H_{24}N_4O_3S$	12.61	317.1642	44.0495	0.5	0.5	0.01, —	0.9993	110.5	5.7	103.8	4.6	100.0	1.1
25	Buprofezin	$C_{16}H_{23}N_3OS$	17.42	306.1635	57.0699	0.5	0.5	0.01, —	0.9978	104.4	12.1	106.6	18.6	102.4	3.7
26	Butachlor	$C_{17}H_{26}ClNO_2$	17.52	312.1725	57.0699	0.5	1.0	—, —	0.9988	86.8	17.0	84.9	9.8	102.8	12.1
27	Butamifos	$C_{13}H_{21}N_2O_4PS$	16.50	333.1035	95.9668	0.5	1.0	—, —	0.9984	107.1	10.5	86.0	14.5	106.0	8.8

Table 1. Cont.

NO.	Compound	Formula	RT/Min	Quantitative Ion (m/z)	Production (m/z)	SDL (mg/kg)	LOQ (mg/kg)	MRL (mg/kg; European Union, China)	R^2	1 × LOQ		2 × LOQ		10 × LOQ	
										Rec. (%)	RSD (%)	Rec. (%)	RSD (%)	Rec. (%)	RSD (%)
28	Butylate	$C_{11}H_{23}NOS$	16.72	218.1573	57.0699	10.0	20.0	0.01, —	0.9985	92.2	14.2	72.3	17.8	77.0	6.7
29	Cadusafos	$C_{10}H_{23}O_2PS_2$	14.78	271.0950	96.9508	0.2	0.5	0.01, —	0.9995	73.5	17.7	75.0	11.5	96.0	2.9
30	Carbaryl	$C_{12}H_{11}NO_2$	6.29	202.0863	127.0542	20.0	50.0	0.05, —	0.9952	72.0	13.5	83.0	7.2	88.0	6.2
31	Carbendazim	$C_9H_9N_3O_2$	2.65	192.0768	160.0505	0.1	0.2	0.05, —	0.9992	70.9	12.6	102.9	3.9	107.6	4.6
32	Carbofuran	$C_{12}H_{15}NO_3$	5.87	222.1125	123.0441	0.5	1.0	0.001, —	0.9974	102.3	12.3	115.8	6.1	96.7	11.2
33	Carbofuran-3-Hydroxy	$C_{12}H_{15}NO_4$	3.60	238.1074	107.0491	1.0	1.0	—, —	0.9924	71.0	13.5	99.2	8.5	110.0	13.8
34	Carfentrazone-ethyl	$C_{15}H_{14}Cl_2F_3N_3O_3$	14.29	412.0435	345.9956	1.0	1.0	0.01, —	0.9997	115.7	12.5	92.1	4.0	107.4	15.5
35	Chlorantraniliprole	$C_{18}H_{14}BrCl_2N_5O_2$	8.36	481.9781	283.9216	1.0	1.0	0.05, —	0.9987	97.3	14.9	76.1	11.5	103.3	15.7
36	Chlorfenvinphos	$C_{12}H_{14}Cl_3O_4P$	13.78	358.9768	98.9843	0.5	0.5	0.01, —	0.9990	74.3	19.6	97.8	11.4	90.1	5.0
37	Chloridazon	$C_{10}H_8ClN_3O$	3.67	222.0429	77.0386	0.5	5.0	0.3, —	0.9951	112.5	5.7	105.6	13.2	92.2	13.2
38	Chlormequat	$C_5H_{12}ClN$	0.75	122.0731	58.0651	0.1	0.1	0.5, 0.5	0.9990	118.2	4.9	108.0	3.0	119.3	6.0
39	Chlorotoluron	$C_{10}H_{13}ClN_2O$	6.15	213.0789	72.0449	0.5	0.5	0.01, —	0.9995	98.0	9.7	104.7	5.0	100.2	3.6
40	Chlorpyrifos	$C_9H_{11}Cl_3NO_3PS$	17.76	349.9336	96.9508	5.0	5.0	0.01, —	0.9924	115.4	8.8	95.2	19.9	90.9	19.9
41	Clodinafop-propargyl	$C_{17}H_{13}ClFNO_4$	15.12	350.0590	91.0542	0.5	0.5	—, —	0.9998	116.4	15.1	117.3	8.3	104.1	2.8
42	Clofentezine	$C_{14}H_8Cl_2N_4$	15.40	303.0199	102.0338	10.0	10.0	0.05, —	0.9955	91.8	11.1	83.5	2.5	93.0	4.4
43	Clomazone	$C_{12}H_{14}ClNO_2$	8.00	240.0786	125.0153	2.0	5.0	0.01, —	0.9979	96.6	8.9	94.6	8.7	92.7	8.7
44	Clothianidin	$C_6H_8ClN_5O_2S$	3.54	250.0160	131.9669	2.0	5.0	0.02, —	0.9917	109.8	8.8	101.4	17.2	103.2	17.2
45	Cyanazine	$C_9H_{13}ClN_6$	5.22	241.0963	214.0854	0.5	5.0	—, —	0.9976	106.2	2.6	106.7	16.4	99.2	16.4
46	Cycloate	$C_{11}H_{21}NOS$	15.41	216.1417	55.0542	10.0	20.0	—, —	0.9981	89.2	7.9	84.5	4.0	75.3	4.7
47	Cycloxydim	$C_{17}H_{27}NO_3S$	16.37	326.1784	107.0491	1.0	1.0	0.05, —	0.9994	87.1	15.0	84.6	20.0	91.6	9.6
48	Cyprodinil	$C_{14}H_{15}N_3$	11.76	226.1339	93.0573	0.1	0.5	0.02, —	0.9982	103.3	9.2	104.9	1.7	96.7	2.6
49	Cyromazine	$C_6H_{10}N_6$	0.80	167.1040	85.0509	2.0	2.0	0.01, —	0.9989	73.5	10.6	74.0	9.8	93.1	6.7
50	Desmetryn	$C_8H_{15}N_5S$	5.23	214.1121	172.0651	0.2	0.2	—, —	0.9978	99.7	13.3	90.8	8.9	101.0	3.7
51	Diallate	$C_{10}H_{17}Cl_2NOS$	16.72	270.0481	86.0600	10.0	20.0	—, —	0.9972	93.4	12.7	74.5	3.8	78.1	2.3
52	Diazinon	$C_{12}H_{21}N_2O_3PS$	15.09	305.1083	96.9508	0.2	0.5	0.02, —	0.9984	94.0	7.9	94.0	6.5	94.7	0.9
53	Dichlorvos	$C_4H_7Cl_2O_4P$	5.24	220.9532	109.0049	20.0	20.0	—, —	0.9908	110.3	20.0	81.5	12.2	71.5	14.0
54	Difenoconazole	$C_{19}H_{17}Cl_2N_3O_3$	14.63	406.0720	251.0025	0.5	1.0	0.005, —	0.9979	98.4	6.4	100.1	6.0	101.4	14.3
55	Diflubenzuron	$C_{14}H_9ClF_2N_2O_2$	12.19	311.0393	141.0144	20.0	20.0	0.01, —	0.9954	116.1	14.9	91.0	10.4	90.9	2.1
56	Dimethenamid	$C_{12}H_{18}ClNO_2S$	9.77	276.0820	244.0557	0.2	0.5	0.01, —	0.9970	114.3	13.4	96.9	11.9	91.2	12.7
57	Dimethoate	$C_5H_{12}NO_3PS_2$	3.83	230.0069	198.9647	5.0	5.0	0.01, 0.05	0.9924	94.3	15.2	100.4	18.1	88.1	18.1

Table 1. *Cont.*

NO.	Compound	Formula	RT/Min	Quantitative Ion (m/z)	Production (m/z)	SDL (mg/kg)	LOQ (mg/kg)	MRL (mg/kg; European Union, China)	R^2	1 × LOQ		2 × LOQ		10 × LOQ	
										Rec. (%)	RSD (%)	Rec. (%)	RSD (%)	Rec. (%)	RSD (%)
58	Dimethylvinphos (E)	$C_{10}H_{10}Cl_3O_4P$	11.58	330.9455	127.0155	10.0	20.0	—, —	0.9918	105.9	18.9	93.3	15.0	89.5	4.5
59	Dimethylvinphos (Z)	$C_{10}H_{10}Cl_3O_4P$	10.59	330.9455	127.0155	5.0	5.0	—, —	0.9984	98.8	10.6	94.3	13.4	93.5	13.4
60	Diniconazole	$C_{15}H_{17}Cl_2N_3O$	13.05	326.0821	70.0400	5.0	5.0	0.01, —	0.9980	100.7	3.3	104.6	14.7	94.8	14.7
61	Dinotefuran	$C_7H_{14}N_4O_3$	2.33	203.1139	58.0526	5.0	10.0	0.1, —	0.9975	81.0	19.3	106.5	3.5	96.2	6.5
62	Dioxabenzofos	$C_8H_9O_4PS$	9.19	217.0083	77.0386	2.0	5.0	—, —	0.9990	101.7	2.7	98.2	11.8	97.2	11.8
63	Dipropetryn	$C_{11}H_{21}N_5S$	11.42	256.1590	102.0120	0.1	0.5	—, —	0.9995	96.0	5.6	103.4	2.8	98.0	0.5
64	Diuron	$C_9H_{10}Cl_2N_2O$	6.72	233.0243	72.0449	0.5	0.5	0.05, —	1.0000	97.4	8.8	92.4	5.3	103.3	2.2
65	Edifenphos	$C_{14}H_{15}O_2PS_2$	13.54	311.0324	109.0107	0.5	0.5	—, —	0.9981	104.8	7.0	101.9	1.9	104.4	1.4
66	Emamectin B1a	$C_{49}H_{75}NO_{13}$	15.63	886.5311	158.1176	0.2	0.5	0.01, —	0.9980	92.6	9.3	113.7	14.2	93.9	3.5
67	Ethion	$C_9H_{22}O_4P_2S_4$	17.97	384.9949	199.0011	1.0	1.0	0.01, —	0.9970	111.4	14.9	107.3	16.2	101.1	11.7
68	Ethoprophos	$C_8H_{19}O_2PS_2$	10.96	243.0637	96.9508	0.5	0.5	0.01, —	0.9991	91.6	17.4	88.0	6.0	93.4	2.8
69	Etrimfos	$C_{10}H_{17}N_2O_4PS$	14.61	293.0719	124.9821	0.5	1.0	—, —	0.9986	114.6	5.3	107.6	7.6	96.4	7.7
70	Fenamidone	$C_{17}H_{17}N_3OS$	10.94	312.1165	92.0495	0.5	0.5	0.01, —	0.9957	82.7	16.2	110.9	8.1	103.7	3.2
71	Fenamiphos	$C_{13}H_{22}NO_3PS$	10.60	304.1131	201.9848	0.5	0.5	0.005, —	0.9979	100.2	7.4	91.1	5.7	100.2	2.2
72	Fenamiphos-sulfone	$C_{13}H_{22}NO_5PS$	5.65	336.1029	266.0247	0.2	0.5	—, —	0.9988	111.4	5.2	93.3	5.6	100.8	3.5
73	Fenamiphos-sulfoxide	$C_{13}H_{22}NO_4PS$	4.65	320.1080	108.0573	0.1	0.5	—, —	0.9988	94.9	7.4	97.4	2.5	101.0	1.4
74	Fenarimol	$C_{17}H_{12}Cl_2N_2O$	10.69	331.0399	81.0447	1.0	5.0	0.02, —	0.9980	97.2	2.0	104.1	11.9	101.2	11.9
75	Fenbuconazole	$C_{19}H_{17}ClN_4$	12.50	337.1215	70.0400	1.0	1.0	0.05, —	0.9992	77.7	4.9	86.2	11.5	107.7	10.3
76	Fenobucarb	$C_{12}H_{17}NO_2$	8.91	208.1332	77.0386	20.0	20.0	—, —	0.9906	88.2	16.7	87.5	11.2	89.9	1.0
77	Fensulfothion	$C_{11}H_{17}O_4PS_2$	7.53	309.0379	140.0290	0.5	0.5	—, —	0.9986	99.8	4.1	115.2	6.9	101.1	1.6
78	Fenthion-sulfoxide	$C_{10}H_{15}O_4PS_2$	6.06	295.0222	109.0049	0.2	0.5	—, —	0.9982	103.6	8.1	100.5	4.3	98.5	1.6
79	Fluacrypyrim	$C_{20}H_{21}F_3N_2O_5$	16.71	427.1475	145.0648	0.5	0.1	—, —	0.9992	92.6	15.9	104.3	6.9	101.5	3.1
80	Fluazifop-butyl	$C_{19}H_{20}F_3NO_4$	17.73	384.1417	91.0542	0.1	0.1	—, —	0.9974	113.1	11.1	107.3	9.5	117.5	16.4
81	Flubendiamide	$C_{23}H_{22}F_7IN_2O_4S$	14.68	705.0125	530.9799	0.2	0.5	0.1, —	0.9987	106.8	2.8	97.7	5.6	99.6	2.8
82	Flumiclorac-pentyl	$C_{21}H_{23}ClFNO_5$	17.51	441.1593	308.0484	0.5	1.0	—, —	0.9963	109.9	11.3	97.6	13.7	81.7	16.8
83	Fluopicolide	$C_{14}H_8Cl_3F_3N_2O$	11.97	382.9727	172.9556	1.0	1.0	0.02, —	0.9991	90.2	10.1	101.6	4.8	104.8	12.3
84	Fluquinconazole	$C_{16}H_8Cl_2FN_5O$	11.52	376.0163	306.9836	10.0	10.0	0.01, —	0.9988	94.2	14.9	95.3	4.5	95.0	1.8
85	Fluridone	$C_{19}H_{14}F_3NO$	9.35	330.1100	309.0960	0.1	0.1	—, —	0.9988	114.7	11.4	95.3	5.9	102.1	1.9
86	Flusilazole	$C_{16}H_{15}F_2N_3Si$	12.45	316.1076	247.0749	0.5	1.0	0.02, —	0.9974	114.7	7.7	93.4	2.6	102.6	12.7
87	Flutriafol	$C_{16}H_{13}F_2N_3O$	6.46	302.1099	70.0400	0.5	1.0	0.01, —	0.9979	99.2	3.9	100.4	3.2	102.6	15.4

Table 1. Cont.

NO.	Compound	Formula	RT/Min	Quantitative Ion (m/z)	Production (m/z)	SDL (mg/kg)	LOQ (mg/kg)	MRL (mg/kg) European Union, China	R^2	1 × LOQ Rec. (%)	1 × LOQ RSD (%)	2 × LOQ Rec. (%)	2 × LOQ RSD (%)	10 × LOQ Rec. (%)	10 × LOQ RSD (%)
88	Fluxapyroxad	$C_{18}H_{12}F_5N_3O$	11.58	382.0973	342.0849	0.5	0.5	0.02, —	0.9995	119.7	11.5	110.1	3.3	95.5	2.5
89	Fonofos	$C_{10}H_{15}OPS_2$	15.40	247.0375	80.9558	5.0	10.0	—, —	0.9960	116.8	7.3	103.8	6.7	89.9	3.7
90	Fosthiazate	$C_9H_{18}NO_3PS_2$	6.44	284.0538	104.0165	0.5	0.5	—, —	0.9992	118.5	6.3	98.8	7.1	94.4	4.1
91	Furathiocarb	$C_{18}H_{26}N_2O_5S$	17.31	383.1635	195.0474	0.1	0.5	0.001, —	0.9987	95.2	9.1	100.6	4.1	103.9	2.1
92	Haloxyfop	$C_{15}H_{11}ClF_3NO_4$	12.37	362.0401	316.0347	20.0	20.0	0.015, —	0.9972	79.2	10.0	103.2	4.5	86.8	3.3
93	Haloxyfop-2-ethoxyethyl	$C_{19}H_{19}ClF_3NO_5$	17.12	434.0977	91.0542	0.5	0.5	—, —	0.9986	116.0	12.2	117.6	8.1	101.0	2.1
94	Haloxyfop-methyl	$C_{16}H_{13}ClF_3NO_4$	16.30	376.0546	272.0085	0.5	0.5	—, —	0.9993	93.1	17.6	111.8	8.1	100.4	2.5
95	Hexaconazole	$C_{14}H_{17}Cl_2N_3O$	12.29	314.0825	70.0400	1.0	5.0	0.05, —	0.9972	91.6	3.3	103.8	12.3	97.3	12.3
96	Hexythiazox	$C_{17}H_{21}ClN_2O_2S$	17.76	353.1085	168.0570	5.0	5.0	0.02, —	0.9987	117.6	8.2	99.6	10.3	90.7	10.3
97	Imazalil	$C_{14}H_{14}Cl_2N_2O$	5.78	297.0550	69.0447	0.2	0.5	0.01, —	0.9979	98.6	15.5	112.1	11.9	99.2	1.8
98	Imazapyr	$C_{13}H_{15}N_3O_3$	3.11	262.1186	69.0699	1.0	5.0	0.01, —	0.9987	102.0	2.4	98.6	17.2	93.2	17.2
99	Imidacloprid	$C_9H_{10}ClN_5O_2$	3.73	256.0596	209.0589	10.0	10.0	—, —	0.9908	105.4	17.2	101.5	7.9	88.4	7.5
100	Imidacloprid-Olefin	$C_9H_8ClN_5O_2$	3.07	254.0439	171.0665	5.0	5.0	—, —	0.9948	115.6	10.6	113.0	12.9	98.7	12.9
101	Iprobenfos	$C_{13}H_{21}O_3PS$	12.40	289.1022	91.0542	5.0	5.0	0.01, —	0.9985	108.2	16.2	100.3	11.6	88.9	11.6
102	Iprovalicarb	$C_{18}H_{28}N_2O_3$	10.60	321.2173	119.0855	1.0	1.0	—, —	0.9987	118.7	12.8	95.6	7.4	101.5	13.5
103	Isazofos	$C_9H_{17}ClN_3O_3PS$	13.69	314.0490	119.9957	0.1	0.5	0.01, —	0.9976	108.4	5.5	106.6	3.9	99.3	2.8
104	Isofenphos	$C_{15}H_{24}NO_4PS$	16.54	346.1236	121.0287	20.0	20.0	—, —	0.9973	113.5	10.4	107.3	15.7	94.7	5.0
105	Isoproturon	$C_{12}H_{18}N_2O$	6.73	207.1492	72.0444	0.2	0.5	0.01, —	0.9995	100.0	9.3	100.4	3.5	103.2	1.5
106	Isopyrazam	$C_{20}H_{23}F_2N_3O$	15.74	360.1895	320.1758	0.5	0.5	0.01, —	0.9979	105.6	9.5	105.0	3.4	97.9	0.9
107	Kresoxim-methyl	$C_{18}H_{19}NO_4$	14.39	314.1387	116.0495	5.0	5.0	0.02, —	0.9991	82.8	12.7	105.8	7.1	98.1	7.1
108	Linuron	$C_9H_{10}Cl_2N_2O_2$	9.22	249.0192	132.9606	5.0	5.0	0.01, —	0.9986	102.3	14.3	97.5	11.1	95.0	11.1
109	Malaoxon	$C_{10}H_{19}O_7PS$	5.77	315.0662	99.0077	0.1	0.5	0.02, —	0.9984	116.8	7.6	97.2	4.6	97.9	1.8
110	Malathion	$C_{10}H_{19}O_6PS_2$	12.60	331.0433	99.0077	1.0	1.0	0.02, —	0.9995	119.3	16.3	104.4	7.1	103.0	12.0
111	Mepanipyrim	$C_{14}H_{13}N_3$	11.59	224.1182	77.0386	0.5	5.0	0.01, —	0.9984	98.6	4.1	109.0	11.9	98.1	11.9
112	Metaflumizone	$C_{24}H_{16}F_6N_4O_2$	17.44	507.1250	178.0463	10.0	10.0	0.01, —	0.9973	105.7	18.1	95.4	15.6	91.9	6.6
113	Metalaxyl	$C_{15}H_{21}NO_4$	6.76	280.1543	45.0335	0.1	0.2	0.01, —	0.9995	105.1	10.5	118.3	12.0	103.6	3.3
114	Metconazole	$C_{17}H_{22}ClN_3O$	12.54	320.1524	70.0400	5.0	5.0	0.02, —	0.9974	102.2	2.1	101.4	15.0	98.7	15.0
115	Methiocarb	$C_{11}H_{15}NO_2S$	8.96	226.0896	121.0648	10.0	50.0	0.03, —	0.9943	72.0	6.6	78.0	5.8	89.0	5.1
116	Methiocarb-sulfoxide	$C_{11}H_{15}NO_3S$	3.51	242.0845	122.0726	0.5	0.5	0.03, —	0.9945	98.8	13.7	94.0	6.3	114.4	6.4
117	Metolachlor	$C_{15}H_{22}ClNO_2$	12.41	284.1412	252.1150	0.2	0.5	0.01, —	0.9987	87.7	10.6	116.9	7.4	97.4	3.0

Table 1. Cont.

NO.	Compound	Formula	RT/Min	Quantitative Ion (m/z)	Production (m/z)	SDL (mg/kg)	LOQ (mg/kg)	MRL (mg/kg; European Union, China)	R^2	1 × LOQ		2 × LOQ		10 × LOQ	
										Rec. (%)	RSD (%)	Rec. (%)	RSD (%)	Rec. (%)	RSD (%)
118	Metrafenone	$C_{19}H_{21}BrO_5$	16.32	409.0645	209.0808	0.2	0.5	0.01, —	0.9989	111.0	13.2	112.2	13.5	99.1	2.7
119	Metribuzin	$C_8H_{14}N_4OS$	5.33	215.0961	49.0106	1.0	5.0	0.1, —	0.9975	99.4	2.4	99.0	11.9	98.8	11.9
120	Mevinphos	$C_7H_{13}O_6P$	3.43	225.0523	127.0155	2.0	5.0	—, —	0.9919	74.4	19.6	112.2	16.5	74.5	16.5
121	Monocrotophos	$C_7H_{14}NO_5P$	2.81	224.0682	58.0287	0.5	0.5	—, —	0.9986	74.0	17.7	80.6	18.3	105.2	8.3
122	Myclobutanil	$C_{15}H_{17}ClN_4$	10.67	289.1215	70.0400	5.0	5.0	0.01, —	0.9993	103.9	7.8	105.0	14.4	94.1	14.4
123	Napropamide	$C_{17}H_{21}NO_2$	11.72	272.1645	171.0804	0.2	0.5	0.01, —	0.9985	105.3	12.7	113.0	5.5	98.0	1.2
124	Norflurazon	$C_{12}H_9ClF_3N_3O$	7.15	304.0459	140.0306	0.1	0.2	—, —	0.9977	92.7	8.1	94.2	4.7	96.4	1.1
125	Omethoate	$C_5H_{12}NO_4PS$	2.10	214.0297	182.9875	0.5	0.5	0.01, —	0.9993	101.0	8.6	104.0	5.1	99.1	3.2
126	Oxadixyl	$C_{14}H_{18}N_2O_4$	5.06	279.1339	132.0808	1.0	1.0	0.01, —	0.9968	101.5	12.7	98.6	8.7	103.1	12.9
127	Paclobutrazol	$C_{15}H_{20}ClN_3O$	8.77	294.1368	70.0400	1.0	1.0	0.01, —	0.9993	94.4	7.4	93.5	3.4	106.5	13.8
128	Pendimethalin	$C_{13}H_{19}N_3O_4$	17.75	282.1448	92.0495	10.0	20.0	0.02, —	0.9963	102.6	10.8	108.9	9.2	81.1	5.6
129	Penthiopyrad	$C_{16}H_{20}F_3N_3OS$	14.57	360.1362	256.0351	0.2	0.5	0.01, —	0.9979	113.8	9.1	101.5	5.7	100.2	3.0
130	Phenthoate	$C_{12}H_{17}O_4PS_2$	15.02	321.0379	79.0542	5.0	20.0	—, —	0.9938	97.1	11.2	88.6	4.9	82.2	2.8
131	Phorate-Sulfone	$C_7H_{17}O_4PS_3$	8.65	293.0097	96.9508	20.0	20.0	0.01, —	0.9982	96.0	13.2	113.8	5.7	82.5	4.0
132	Phorate-sulfoxide	$C_7H_{17}O_3PS_3$	6.37	277.0150	96.9508	0.5	0.5	0.01, —	0.9992	105.4	9.4	97.6	6.5	109.1	3.4
133	Phosalone	$C_{12}H_{15}ClNO_4PS_2$	16.04	367.9941	110.9996	20.0	20.0	0.01, —	0.9990	119.9	15.3	109.9	5.7	86.9	5.6
134	Phosphamidon	$C_{10}H_{19}ClNO_5P$	4.73	300.0762	127.0155	0.2	0.5	—, —	0.9978	95.1	4.2	95.5	4.3	104.1	2.2
135	Phoxim	$C_{12}H_{15}N_2O_3PS$	16.05	299.0614	77.0389	10.0	20.0	0.02, —	0.9933	90.9	19.6	97.3	4.5	108.2	13.4
136	Picoxystrobin	$C_{18}H_{16}F_3NO_4$	14.80	368.1104	145.0648	0.5	1.0	0.01, —	0.9995	114.8	19.0	71.2	10.8	99.1	16.5
137	Piperonyl Butoxide	$C_{19}H_{30}O_5$	17.12	356.2423	119.0855	0.2	0.5	—, —	0.9993	115.1	16.2	108.4	8.6	100.4	4.6
138	Pirimicarb	$C_{11}H_{18}N_4O_2$	4.42	239.1503	72.0444	0.5	1.0	0.05, —	0.9982	105.8	12.0	95.1	6.5	102.7	14.5
139	Pirimiphos-methyl	$C_{11}H_{20}N_3O_3PS$	15.91	306.1036	164.1182	0.5	0.5	0.01, —	0.9971	104.2	9.3	101.5	5.6	99.7	2.4
140	Pretilachlor	$C_{17}H_{26}ClNO_2$	16.25	312.1725	252.1150	0.2	0.5	—, —	0.9977	98.8	10.0	116.7	6.6	101.2	3.3
141	Prochloraz	$C_{15}H_{16}Cl_3N_3O_2$	13.12	376.0381	70.0287	0.5	0.5	0.03, —	0.9977	115.1	8.7	101.3	8.8	93.7	2.4
142	Profenofos	$C_{11}H_{15}BrClO_3PS$	16.19	372.9424	96.9509	5.0	5.0	0.01, —	0.9989	105.5	13.6	106.3	7.5	97.9	7.5
143	Prometryn	$C_{10}H_{19}N_5S$	8.68	242.1434	68.0243	0.2	0.5	—, —	0.9975	104.2	2.3	101.0	4.1	100.0	1.2
144	Propamocarb	$C_9H_{20}N_2O_2$	2.16	189.1598	74.0237	1.0	1.0	0.01, —	0.9990	88.6	6.1	72.3	6.3	98.0	13.7
145	Propanil	$C_9H_9Cl_2NO$	8.21	218.0134	127.0178	5.0	5.0	0.01, —	0.9954	95.9	13.0	116.3	10.5	91.6	10.5
146	Propaphos	$C_{13}H_{21}O_4PS$	13.19	305.0971	221.0032	0.2	0.5	—, —	0.9987	108.8	9.1	102.8	8.4	96.2	3.4
147	Propargite	$C_{19}H_{26}O_4S$	18.36	368.1886	57.0699	20.0	20.0	0.01, —	0.9906	99.4	10.7	91.2	4.1	82.9	2.8
148	Propazine	$C_9H_{16}ClN_5$	8.22	230.1167	146.0228	0.1	0.1	—, —	0.9972	111.4	6.6	114.9	3.9	96.8	7.4

Table 1. Cont.

NO.	Compound	Formula	RT/Min	Quantitative Ion (m/z)	Production (m/z)	SDL (mg/kg)	LOQ (mg/kg)	MRL (mg/kg; European Union, China)	R^2	1 × LOQ		2 × LOQ		10 × LOQ	
										Rec. (%)	RSD (%)	Rec. (%)	RSD (%)	Rec. (%)	RSD (%)
149	Propiconazole	$C_{15}H_{17}Cl_2N_3O_2$	13.16	342.0771	69.0699	0.1	1.0	0.01, —	0.9982	102.8	6.3	100.0	4.2	102.0	11.7
150	Propyzamide	$C_{12}H_{11}Cl_2NO$	11.12	256.0290	189.9821	5.0	5.0	0.01, —	0.9953	115.0	14.2	98.2	12.5	94.8	12.5
151	Prothioconazole-desthio	$C_{14}H_{15}Cl_2N_3O$	10.55	312.0664	70.0400	0.5	0.5	0.01, —	0.9994	119.4	3.9	108.1	11.1	111.4	1.8
152	Prothiofos	$C_{11}H_{15}Cl_2O_2PS_2$	19.11	344.9701	240.9041	20.0	20.0	—, —	0.9917	118.6	10.6	83.6	10.1	82.0	9.6
153	Pyraclostrobin	$C_{19}H_{18}ClN_3O_4$	15.47	388.1059	194.0812	0.5	0.5	0.01, —	0.9981	119.9	4.3	108.5	6.0	103.0	0.5
154	Pyridaben	$C_{19}H_{25}ClN_2OS$	18.85	365.1449	147.1168	0.5	0.5	0.01, —	0.9969	86.6	3.9	118.6	18.3	104.4	11.6
155	Pyridaphenthion	$C_{14}H_{17}N_2O_4PS$	11.69	341.0719	92.0498	0.5	0.5	—, —	0.9992	100.6	13.8	98.8	19.7	103.5	4.0
156	Pyrimethanil	$C_{12}H_{13}N_3$	7.56	200.1182	77.0386	0.5	0.5	0.05, —	0.9972	102.4	7.0	96.2	4.2	100.1	2.8
157	Pyriproxyfen	$C_{20}H_{19}NO_3$	17.56	322.1438	96.0444	0.5	0.5	0.05, —	0.9977	114.8	13.3	118.8	19.0	108.9	8.2
158	Quinalphos	$C_{12}H_{15}N_2O_3PS$	14.06	299.0614	96.9508	0.5	0.5	—, —	0.9986	113.0	8.7	110.8	3.6	99.1	3.3
159	Quinoxyfen	$C_{15}H_8Cl_2FNO$	16.82	308.0040	196.9789	1.0	1.0	0.05, —	0.9963	117.7	14.5	104.5	15.3	92.9	7.8
160	Quizalofop-ethyl	$C_{19}H_{17}ClN_2O_4$	16.68	373.0950	91.0542	0.5	0.5	—, —	0.9995	99.8	12.5	98.2	11.7	102.0	11.7
161	Saflufenacil	$C_{17}H_{17}ClF_4N_4O_5S$	11.03	501.0617	348.9998	2.0	5.0	0.01, —	0.9982	109.8	8.8	101.4	17.2	103.2	17.2
162	Simazine	$C_7H_{12}ClN_5$	5.04	202.0854	132.0323	0.5	0.5	0.01, —	0.9974	100.8	2.2	99.3	1.7	101.7	1.4
163	Spinosyn A	$C_{41}H_{65}NO_{10}$	12.82	732.4681	142.1226	0.2	0.5	0.2, —	0.9990	100.9	3.5	110.2	6.3	97.6	1.0
164	Spinosyn D	$C_{42}H_{67}NO_{10}$	14.44	746.4838	142.1226	1.0	1.0	0.2, —	0.9991	110.3	13.1	94.1	5.5	99.7	16.4
165	Spirodiclofen	$C_{21}H_{24}Cl_2O_4$	19.01	411.1124	71.0855	1.0	5.0	0.004, —	0.9997	87.0	13.1	106.5	16.8	92.2	16.8
166	Spirotetramat	$C_{21}H_{27}NO_5$	10.19	374.1962	302.1751	5.0	5.0	0.01, —	0.9981	73.0	14.5	96.1	13.1	79.3	13.1
167	Spirotetramat-enol	$C_{18}H_{23}NO_3$	5.33	302.1758	216.1019	0.5	0.5	0.01, —	0.9943	95.1	7.6	118.6	2.5	91.9	5.6
168	Spirotetramat-enol-glucoside	$C_{24}H_{33}NO_8$	2.89	464.2279	302.1751	2.0	5.0	—, —	0.9979	109.8	8.8	101.4	17.2	103.2	17.2
169	Spiroxamine	$C_{18}H_{35}NO_2$	8.31	298.2741	100.1121	0.5	0.5	0.015, —	0.9959	95.4	10.1	105.7	10.5	97.5	2.9
170	Sulfentrazone	$C_{11}H_{10}Cl_2F_2N_4O_3S$	6.43	386.9891	306.9944	5.0	10.0	—, —	0.9987	112.7	15.1	96.7	5.9	96.8	2.0
171	Sulfotep	$C_8H_{20}O_5P_2S_2$	15.80	323.0300	96.9508	1.0	1.0	—, —	0.9970	92.1	4.5	96.0	2.6	92.5	12.0
172	Sulfoxaflor	$C_{10}H_{10}F_3N_3OS$	4.57	278.0569	154.0463	1.0	10.0	0.2, —	0.9989	99.0	7.2	101.5	3.6	92.9	1.3
173	Sulprofos	$C_{12}H_{19}O_2PS_3$	18.03	323.0358	218.9698	5.0	5.0	—, —	0.9990	102.9	18.6	94.4	7.5	91.0	7.5
174	Tebuconazole	$C_{16}H_{22}ClN_3O$	11.84	308.1524	70.0400	1.0	5.0	0.02, —	0.9990	91.7	2.8	103.7	12.9	97.0	12.9
175	Tebufenozide	$C_{22}H_{28}N_2O_2$	14.05	353.2224	133.0648	1.0	10.0	0.01, —	0.9908	118.8	9.3	93.0	10.0	94.9	7.7
176	Terbufos-Sulfone	$C_9H_{21}O_4PS_3$	11.80	321.0412	275.0535	5.0	5.0	0.01, —	0.9992	102.3	9.4	100.0	12.0	101.7	12.0
177	Terbufos-Sulfoxide	$C_9H_{21}O_3PS_3$	8.40	305.0465	130.9385	0.5	1.0	0.01, —	0.9983	109.5	14.5	115.5	8.7	101.4	13.8
178	Terbumeton	$C_{10}H_{19}N_5O$	5.61	226.1662	170.1036	0.2	0.5	—, —	0.9982	92.2	2.2	100.5	3.5	101.9	1.5
179	Terbuthylazine	$C_9H_{16}ClN_5$	8.90	230.1167	174.0541	0.5	0.5	0.02, —	0.9972	111.0	10.5	106.3	11.4	96.4	4.5

Table 1. Cont.

NO.	Compound	Formula	RT/Min	Quantitative Ion (m/z)	Production (m/z)	SDL (mg/kg)	LOQ (mg/kg)	MRL (mg/kg; European Union, China)	R^2	1 × LOQ		2 × LOQ		10 × LOQ	
										Rec. (%)	RSD (%)	Rec. (%)	RSD (%)	Rec. (%)	RSD (%)
180	Terbutryn	$C_{10}H_{19}N_5S$	9.09	242.1434	186.0808	0.2	0.5	—, —	0.9983	99.3	9.0	94.3	6.2	101.9	2.2
181	Tetramethrin	$C_{19}H_{25}NO_4$	17.09	332.1856	164.0706	20.0	20.0	—, —	0.9934	90.9	5.2	89.8	9.4	84.7	4.0
182	Thiabendazole	$C_{10}H_7N_3S$	2.90	202.0433	131.0604	0.2	0.2	0.2, 0.2	0.9996	81.2	16.6	107.1	16.1	72.9	4.3
183	Thiacloprid	$C_{10}H_9ClN_4S$	4.55	253.0309	126.0087	0.2	0.5	0.05, —	0.9999	106.3	5.7	89.5	7.0	104.8	2.0
184	Thiamethoxam	$C_8H_{10}ClN_5O_3S$	3.17	292.0266	131.9664	0.5	1.0	0.05, —	0.9994	100.8	16.1	91.4	9.6	103.4	17.3
185	Thiobencarb	$C_{12}H_{16}ClNO5$	15.23	258.0714	125.0153	1.0	5.0	0.01, —	0.9985	100.0	5.2	99.4	8.2	92.6	8.2
186	Thiophanate-methyl	$C_{12}H_{14}N_4O_4S_2$	5.50	343.0529	151.0324	2.0	20.0	0.05, —	0.9995	78.2	13.8	81.7	4.4	77.7	14.2
187	Tolfenpyrad	$C_{21}H_{22}ClN_3O_2$	16.96	384.1477	197.0961	0.5	0.5	—, —	0.9994	108.7	16.7	118.0	12.1	102.2	7.5
188	Triadimefon	$C_{14}H_{16}ClN_3O_2$	11.26	294.1004	57.0699	1.0	5.0	0.01, —	0.9993	101.6	3.5	102.4	14.5	97.1	14.5
189	Trichlorfon	$C_4H_8Cl_3O_4P$	3.36	256.9299	78.9945	10.0	10.0	0.01, —	0.9981	105.9	17.6	102.6	13.3	95.8	3.4
190	Trifloxystrobin	$C_{20}H_{19}F_3N_2O_4$	16.78	409.1370	145.0260	0.2	0.5	0.02, —	0.9982	106.0	19.8	109.7	9.2	102.3	1.3
191	Triflumizole	$C_{15}H_{15}ClF_3N_3O$	15.00	346.0929	69.0447	0.5	0.5	0.01, —	0.9980	96.7	11.1	119.4	6.5	99.8	3.1
192	Trinexapac-ethyl	$C_{13}H_{16}O_5$	7.60	253.1071	69.0335	10.0	20.0	—, —	0.9976	73.1	12.7	100.9	4.9	83.2	0.7
193	Uniconazole	$C_{15}H_{18}ClN_3O$	10.67	292.1213	70.0400	0.5	0.5	—, —	0.9980	108.2	11.8	107.6	4.9	101.6	1.3
194	Warfarin	$C_{19}H_{16}O_4$	9.15	309.1121	163.0390	0.5	0.5	0.01, —	0.9991	79.4	19.0	76.6	10.5	92.6	2.3
195	Zoxamide	$C_{14}H_{16}Cl_3NO_2$	15.00	336.0319	186.9712	0.5	0.5	0.01, —	0.9994	95.5	17.6	96.2	6.3	99.8	4.5

RT: retention time; SDL: screening detection limit; LOQ: the limit of quantification; MRL: maximum residue limits; R^2: coefficient of determination. "—" means no MRL value.

2.5. Validation of the Method

The method was validated in the raw milk matrix by evaluating the following parameters: screening detection limit (SDL), the limit of quantification (LOQ), linearity, matrix effect, accuracy, and precision. To define the SDL, refer to the European SANTE/12682/2019 guidelines [28]. LOQs were assessed by determining the lowest concentration of spiked samples where recovery and precision were satisfactory (70–120% and less than 20%, respectively). Calibration curves were investigated by determining the results of a series of standard addition recovery experiments (1–200 µg/kg) of blank matrix extract solutions before injection. Matrix effects were evaluated by comparing the slope of the matrix-matched calibration curve with the solvent calibration curve. To validate the accuracy and precision of the established method, recovery studies were performed for each substrate in six replicates for three spiked levels at $1 \times LOQ$, $2 \times LOQ$, and $10 \times LOQ$.

Agilent Mass Hunter (version B. 08.00) software was used to analyze the data based on the self-built database. To ensure the accuracy of target pesticide identifications, the specific settings of the corresponding screening parameters included the retention time offset threshold (≤ 0.15 min), the co-exist score (≥ 15), the signal-to-noise ratio (≥ 3), the mass deviation (≤ 10 ppm), and the number of characteristic ions in the qualitative identification of compounds (5:2). The data results were analyzed and summarized by Microsoft Excel 2016 (Seattle, WA, USA) software, and the analysis of graphs was drawn by Origin 2018 software.

3. Results

3.1. Optimization of the QuEChERS Procedure

The QuEChERS procedure was evaluated due to the possibility of matrix interferences influencing the identification of compounds, which are the most challenging situations in high-throughput screening and are also required to validate quantitative determination. For this reason, different procedures based on the QuEChERS method have been evaluated as follows.

3.1.1. Optimization of the Extraction Solvent Volume

This study used acetonitrile with 1% acetate as an extraction solvent because it can extract various compounds with different polarity ranges and is the most effective organic solvent in multi-residue methods [17,18,20]. The volumes of extraction solution, such as 10 mL, 16 mL, and 20 mL of acetonitrile with 1% acetate, were compared to improve the extraction efficiency. In the spiked level of 100 µg/kg, the detected pesticides were 170, 173, and 166, respectively, using 10 mL, 16 mL, and 20 mL of acetonitrile with 1% acetate for raw milk. By 10 mL of the extraction solution, the final sample solution contains a high matrix background interference, affecting the definitive identification of compounds under the same purification conditions. Moreover, when the extraction solution volume was 20 mL, the sample solution was diluted by a factor of five, which noticeably reduced the sensitivity of the compound detection. Ultimately, the relatively good experimental results could be found when the volume of the extraction solution was 16 mL. Considering the response of the target pesticide and background interference, 16 mL acetonitrile with 1% acetate was selected for the extraction solvent.

3.1.2. Optimization of the Type of Extraction Salt

The matrix environment, especially pH, may play an essential role in extracting some pesticides during the extraction process. Therefore, the effect of pH on pesticide recovery has been frequently investigated in many studies [27]. Extraction salts could adjust the pH of the matrix and affect the extraction efficiency by reducing the solubility of the target pesticides in an aqueous solution and enhancing their transfer into the extraction solution. To assess the extraction salt, the various compositions of salt pocket from the initial method (4 g anhydrous $MgSO_4$ and 1 g sodium chloride), the AOAC method (6 g anhydrous $MgSO_4$ and 1.5 g sodium acetate), and the EN method (4 g anhydrous $MgSO_4$, 1 g anhydrous NaCl,

1 g dihydrate trisodium citrate, and 0.5 g disodium citrate) [29] were compared. As shown in Figure 1, the number of pesticides with the recovery in 70–120% by the EN method was slightly higher than the other two methods. This is because citrate buffering (EN) gently adjusts the pH of the matrix to between 5.0 and 5.5, enabling the satisfactory recovery of some sensitive pesticides under acidic or basic conditions. The results also verified that pH-sensitive pesticides, such as carbofuran and carbofuran-3-hydroxy (carbamate pesticides), had good performance and stability effects through EN buffer salts. Therefore, the EN method salt pocket was selected.

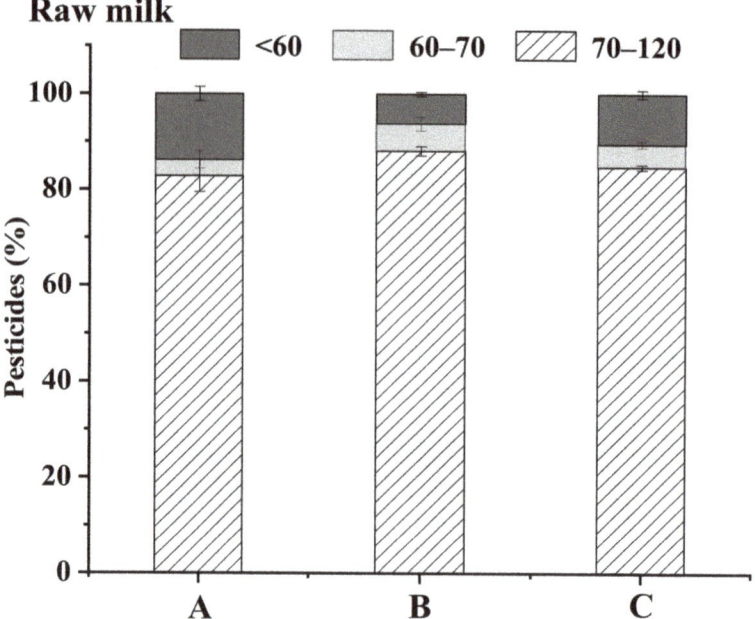

Figure 1. Recoveries (%) obtained for various salt pockets methods; (**A**) 4 g anhydrous MgSO$_4$, 1 g sodium chloride, (**B**) 4 g anhydrous MgSO$_4$, 1 g anhydrous NaCl, 1 g dihydrate trisodium citrate and 0.5 g disodium citrate, and (**C**) 6 g anhydrous MgSO$_4$, 1.5 g sodium acetate.

3.1.3. Optimization of the Freezing Temperature

The low-temperature precipitation step enables the removal of a large proportion of interfering substances, such as lipids, fats, and proteins that may be extracted along with the target pesticide residues. The significant advantage of this purification technology is that it is simple to operate and does not require specialized equipment [30]. The main components of milk are protein and animal oil esters. Therefore, it was necessary to use a low-temperature precipitation method for the raw milk to reduce the co-extracts in the extracts. As shown in Figure 2, the TIC chromatograms of different experimental groups overlapped, indicating a significant reduction in the signal intensity of co-extractives and matrix-derived interferences under low-temperature conditions. Meanwhile, the results showed that the recovery and precision of pesticides frozen at −20 °C for 0.5 h were better than those of the experimental group without freezing. Still, the results were similar to those of the experimental group frozen for 1.0 h. Thus, a freezing time of 0.5 h was chosen in the final method.

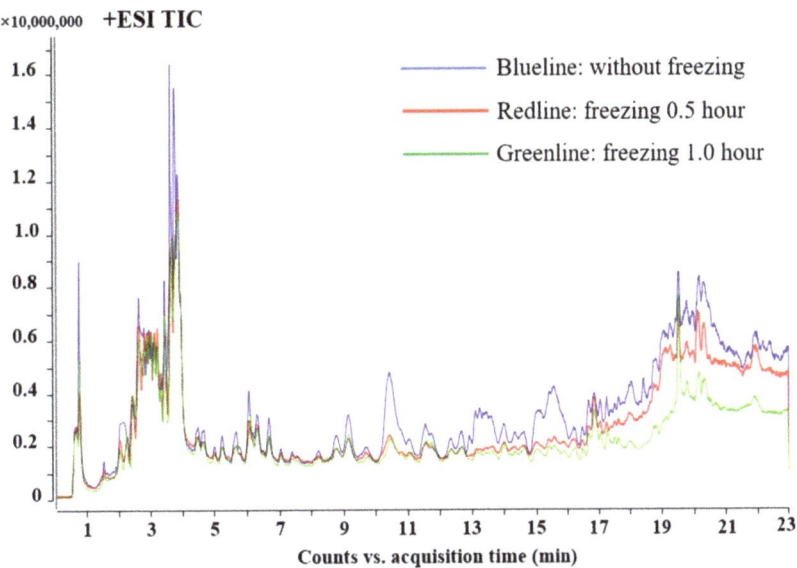

Figure 2. LC-Q-TOF/MS Total ion chromatogram overlap showing the effect of freezing (Blueline: without freezing; Redline: freezing 0.5 h; Greenline: freezing 1.0 h).

3.1.4. Optimization of the Purification Adsorbent

Despite the sample solution being frozen-out to remove most of the interfering substances, the remaining matrix components may still interfere with the determination and contaminate the LC-Q-TOF/MS system, so it is necessary to develop an additional efficient clean-up step. Sorbents play a crucial role in the QuEChERS method. Various sorbents such as primary secondary amines (PSA) and octadecyl (C18) are often used for sample clean-up in pesticide residue analysis. C18 is a reversed-phase adsorption material that removes non-polar impurities such as lipids, cholesterol, and lipophilic compounds. PSA is a weak anion exchange sorbent that could adsorb polar molecules and effectively remove co-extracted components from the matrix, such as organic acids and sugars [27].

Raw milk is a complicated matrix with high lipid, fat, and protein intensities. Thus, the optimization of the purification step is achieved by different adsorbent combinations and dosage variables. In the present experiment, 500 mg of anhydrous magnesium sulfate was applied to remove the residual water. In addition, five different types of sorbents (100 mg of C18, 200 mg of C18, 300 mg of C18, 50 mg of PSA, and 50 mg of PSA + 200 mg of C18) were tested to investigate the influences on recoveries in raw milk.

According to SANTE/12682/2019 guidelines, the acceptable recovery interval is 70–120%, with an RSD less than or equal to 20% for multi-residue methods. As shown in Figure 3, the most significant number of pesticides with satisfactory recoveries and RSDs were found when 200 mg of C18 was used, along with better peak shapes and less matrix interference for some drugs, such as thiophanate-methyl. It may be that 200 mg of C18 can remove more interfering substances without affecting the pesticide detection, but excessive use of C18 will adsorb pesticides to reduce the recovery. Meanwhile, PSA adsorbent alone could not effectively remove lipids and proteins, which affected the detection of target pesticides. Finally, based on these results, 200 mg of C18 was selected as the sorbent to clean-up raw milk samples in this study.

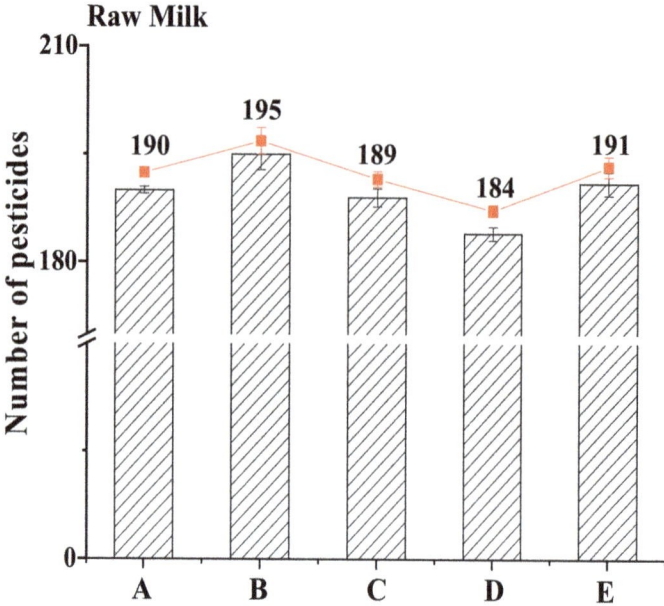

Figure 3. Comparison of different sorbents for dispersive-SPE clean-up of analytes in raw milk. (**A**): 100 mg C18; (**B**): 200 mg C18; (**C**): 300 mg C18; (**D**): 50 mg PSA; and (**E**): 50 mg PSA+ 200 mg C18.

3.2. Matrix Effect

The co-eluting components, such as lipids, fats, and proteins in raw milk interfere with the ionization of pesticides with the suppression or the enhancement of the response. The formula evaluated the matrix effect in raw milk: the matrix effect (ME, %) = (slope of the matrix standard curve/slope of the solvent standard curve − 1) × 100. Matrix effects can be classified into three categories based on the results of the calculated data (Strong matrix effect: $|ME| \geq 50$; Medium matrix effect: $20 < |ME| < 50$; and Weak matrix effect: $|ME| \leq 20$) [23]. As shown in Figure 4, more than 89.2% of the pesticides had a weak matrix effect in raw milk. The data results indicate that the method accurately analyzes trace pesticide residues in milk.

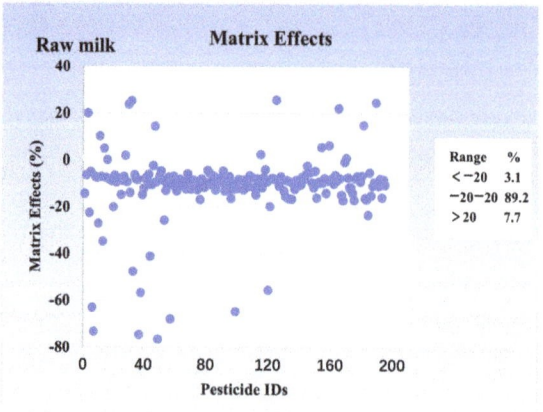

Figure 4. Matrix effect distribution of pesticides in raw milk analysis methods.

3.3. Method Validation

The linearity, SDL, LOQ, accuracy, and precision were determined to evaluate the performance of the modified QuEChERS method. The linearity was selected in the 1–200 µg/kg concentration range. As presented in Table 1, the coefficients of determination (R^2) were higher than 0.99 for the pesticides in different linear ranges.

The sensitivity of the method was performed by SDL according to SANTE/12682/2019. SDLs were determined by spiking a series of mixed standard solutions in 20 blank samples and the lowest level at which pesticides had been screened in at least 95% of the samples [28]. As shown in Figure 5A, the percentage of pesticides with SDLs no more than 10 µg/kg was 93.3% for raw milk. LOQs were determined as the lowest validated spike level based on the recovery results by spiking a series of mixed standard solutions in blank samples. For raw milk, the LOQs were in the range of 0.5–50 µg/kg, and more than 87.2% of pesticides were less than or equal to 10 µg/kg, as shown in Figure 5B. The details of the SDLs and LOQs are listed in Table 1.

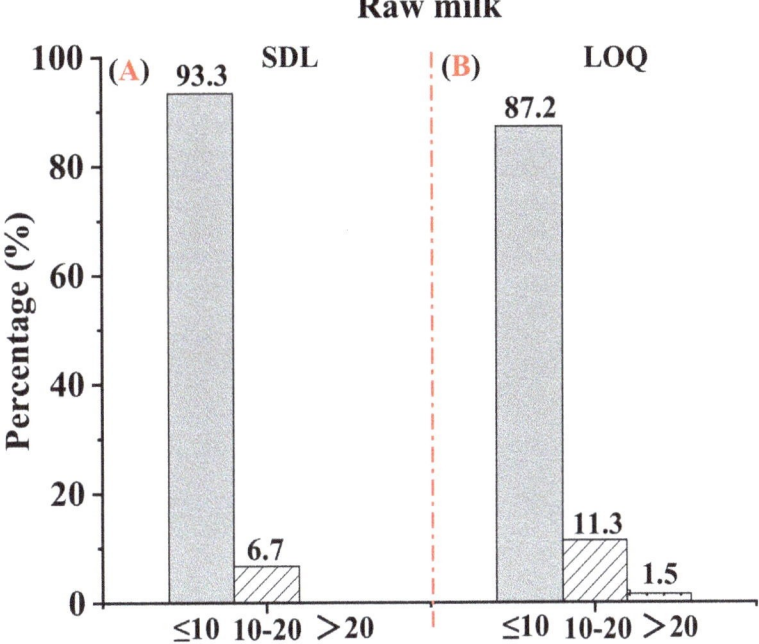

Figure 5. The distribution of the screening and quantification limits of pesticides in raw milk: (**A**) SDL distribution of pesticides in raw milk; (**B**) LOQ distribution of pesticides in raw milk.

For the accuracy and precision assessment, six replicates at three spiked levels were used, including 1 × LOQ, 2 × LOQ, and 10 × LOQ. The overall accuracy values for quantifying target pesticides in raw milk through recovery experiments ranged between 70.0% and 119.8%. The lowest accuracy value was relative to aminopyralid (70.0%). Thus, the method's precision can be considered appropriate (SANTE/12682/2019). For 195 pesticide residues, the RSD values ranged from 0.5 to 20.0% under in-laboratory conditions in all recovery experiments, indicating that the method's precision was acceptable. Therefore, it could be concluded that the modified QuEChERS method was sufficiently sensitive to determine the residues of the investigated pesticides in raw milk samples. The experimental results of the method performance evaluation, including recovery values (Rec, %) and RSD (%), are shown in Table 1.

3.4. Analysis of Real Samples

The established method was applied to 21 actual raw milk samples collected from local dairy farms in China (six batches from the Inner Mongolia Autonomous Region, six batches from Shaanxi Province, six batches from Shandong Province, and three batches from Hebei Province). Raw milk samples were collected at the dairy farm, transported to the laboratory using the cold chain, and stored at −20 °C. Samples need to be thawed to room temperature before analysis. To guarantee the accuracy and reliability of the experimental results, the spiked samples were tested simultaneously. The samples were pretreated according to the preparation section and then analyzed by LC-Q-TOF/MS. The results obtained showed that no pesticides were detected in the actual samples. The recovery results of the quality control samples met the analytical requirements, indicating that the values were accurate and reliable.

4. Conclusions

A high-throughput screening method based on modified QuEChERS and LC-Q-TOF/MS was established to analyze multi-residue pesticides in raw milk rapidly. The modified QuEChERS sample preparation method used an EN salting agent, followed by a freezing treatment, and then a purification treatment with C18 adsorbent, which effectively removed interference and reduced the matrix effect of multiple pesticide residues in raw milk. Overall, 195 pesticides passed the validation with satisfactory recoveries (70−120%) and an RSD of ≤20%. The method exhibited a good sensitivity to milk matrices, and the percentage of pesticides with SDL and LOQ values not exceeding 10 µg/kg for the established method were 93.3% and 87.2%, respectively. These results show that the method is cost-effective, convenient, and reliable for the routine screening of pesticide residues in raw milk and fully complies with the requirements of relevant regulations.

Author Contributions: Conceptualization, X.W.; Data curation, K.T.; Formal analysis, X.W., K.T. and Y.X.; Investigation, Y.X.; Methodology, H.C.; Project administration, C.F.; Resources, C.Y., S.H. and W.W.; Software, K.T., M.L. and W.W.; Supervision, C.F. and H.C.; Validation, C.Y. and M.L.; Writing—original draft, X.W.; Writing—review & editing, H.C. All authors have read and agreed to the published version of the manuscript.

Funding: This work was financially supported by the Science and Technology Project of the State Administration for Market Regulation (2021MK165).

Institutional Review Board Statement: Not applicable.

Informed Consent Statement: Not applicable.

Data Availability Statement: Not applicable.

Conflicts of Interest: The authors declare no conflict of interest.

References

1. Givens, D. MILK Symposium review: The importance of milk and dairy foods in the diets of infants, adolescents, pregnant women, adults, and the elderly. *J. Dairy Sci.* **2020**, *103*, 9681–9699. [CrossRef] [PubMed]
2. Sheng, F.; Wang, J.; Chen, K.Z.; Fan, S.; Gao, H. Changing Chinese Diets to Achieve a Win–Win Solution for Health and the Environment. *China World Econ.* **2021**, *29*, 34–52. [CrossRef]
3. Liu, L.; Wang, Y.; Ariyawardana, A. Rebuilding milk safety trust in China: What do we learn and the way forward. *J. Chin. Gov.* **2021**, *6*, 1–23. [CrossRef]
4. Gill, J.P.S.; Bedi, J.S.; Singh, R.; Fairoze, M.N.; Hazarika, R.A.; Gaurav, A.; Satpathy, S.K.; Chauhan, A.S.; Lindahl, J.; Grace, D.; et al. Pesticide Residues in Peri-Urban Bovine Milk from India and Risk Assessment: A Multicenter Study. *Sci. Rep.* **2020**, *10*, 1–11. [CrossRef]
5. Tsakiris, I.N.; Goumenou, M.; Tzatzarakis, M.N.; Alegakis, A.K.; Tsitsimpikou, C.; Ozcagli, E.; Tsatsakis, A.M. Risk assessment for children exposed to DDT residues in various milk types from the Greek market. *Food. Chem. Toxicol.* **2015**, *75*, 156–165. [CrossRef]
6. Lachat, L.; Glauser, G. Development and Validation of an Ultra-Sensitive UHPLC–MS/MS Method for Neonicotinoid Analysis in Milk. *J. Agric. Food Chem.* **2018**, *66*, 8639–8646. [CrossRef]
7. LeDoux, M. Analytical methods applied to the determination of pesticide residues in foods of animal origin. A review of the past two decades. *J. Chromatogr. A* **2011**, *1218*, 1021–1036. [CrossRef]

8. Năstăsescu, V.; Mititelu, M.; Goumenou, M.; Docea, A.O.; Renieri, E.; Udeanu, D.I.; Oprea, E.; Arsene, A.L.; Dinu-Pîrvu, C.E.; Ghica, M. Heavy metal and pesticide levels in dairy products: Evaluation of human health risk. *Food Chem. Toxicol.* **2020**, *146*, 111844. [CrossRef]
9. Ramezani, S.; Mahdavi, V.; Gordan, H.; Rezadoost, H.; Conti, G.O.; Khaneghah, A.M. Determination of multi-class pesticides residues of cow and human milk samples from Iran using UHPLC-MS/MS and GC-ECD: A probabilistic health risk assessment. *Environ. Res.* **2022**, *208*, 112730. [CrossRef]
10. European Commission. Pesticide Residue Online Database in/on Milk. Available online: https://ec.europa.eu/food/plant/pesticides/eu-pesticides-database/mrls/?event=search.pr (accessed on 15 March 2022).
11. *GB 2763-2021*; National Food Safety Standard-In Maximum Residue Limits for Pesticides in Food. China Agriculture Press: Beijing, China, 2021.
12. Rejczak, T.; Tuzimski, T. QuEChERS-based extraction with dispersive solid phase extraction clean-up using PSA and ZrO2-based sorbents for determination of pesticides in bovine milk samples by HPLC-DAD. *Food Chem.* **2017**, *217*, 225–233. [CrossRef]
13. Tripathy, V.; Sharma, K.K.; Yadav, R.; Devi, S.; Tayade, A.; Sharma, K.; Shakil, N.A. Development, validation of QuEChERS-based method for simultaneous determination of multiclass pesticide residue in milk, and evaluation of the matrix effect. *J. Environ. Sci. Health B* **2019**, *54*, 394–406. [CrossRef]
14. Manav, Ö.G.; Dinç-Zor, Ş.; Alpdoğan, G. Optimization of a modified QuEChERS method by means of experimental design for multiresidue determination of pesticides in milk and dairy products by GC–MS. *Microchem. J.* **2019**, *144*, 124–129. [CrossRef]
15. Zheng, G.; Han, C.; Liu, Y.; Wang, J.; Zhu, M.; Wang, C.; Shen, Y. Multiresidue analysis of 30 organochlorine pesticides in milk and milk powder by gel permeation chromatography-solid phase extraction-gas chromatography-tandem mass spectrometry. *J. Dairy Sci.* **2014**, *97*, 6016–6026. [CrossRef]
16. Kang, H.S.; Kim, M.; Kim, E.J.; Choe, W.-J. Determination of 66 pesticide residues in livestock products using QuEChERS and GC–MS/MS. *Food Sci. Biotechnol.* **2020**, *29*, 1573–1586. [CrossRef]
17. Imamoglu, H.; Oktem Olgun, E. Analysis of veterinary drug and pesticide residues using the ethyl acetate multiclass/multiresidue method in milk by liquid chromatography-tandem mass spectrometry. *J. Anal. Method Chem.* **2016**, *2016*, 2170165. [CrossRef]
18. Görel-Manav, Ö.; Dinç-Zor, Ş.; Akyildiz, E.; Alpdoğan, G. Multivariate optimization of a new LC–MS/MS method for the determination of 156 pesticide residues in milk and dairy products. *J. Sci. Food Agric.* **2020**, *100*, 4808–4817. [CrossRef]
19. Jadhav, M.R.; Pudale, A.; Raut, P.; Utture, S.; Shabeer, T.A.; Banerjee, K. A unified approach for high-throughput quantitative analysis of the residues of multi-class veterinary drugs and pesticides in bovine milk using LC-MS/MS and GC–MS/MS. *Food Chem.* **2019**, *272*, 292–305. [CrossRef]
20. Jia, W.; Zhang, R.; Shi, L.; Zhang, F.; Xu, X.; Chu, X. Construction of Non-Target Screening Method for Pesticides in Milk and Dairy Products Based on Mass Spectrometry Fracture Mechanism. *Chin. J. Anal. Chem.* **2019**, *47*, 1098–1149. [CrossRef]
21. Aydoğan, C.; El Rassi, Z. MWCNT based monolith for the analysis of antibiotics and pesticides in milk and honey by integrated nano-liquid chromatography-high resolution orbitrap mass spectrometry. *Anal. Methods UK* **2019**, *11*, 21–28. [CrossRef]
22. López-Ruiz, R.; Romero-González, R.; Frenich, A.G. Ultrahigh-pressure liquid chromatography-mass spectrometry: An overview of the last decade. *TrAC Trends Anal. Chem.* **2019**, *118*, 170–181. [CrossRef]
23. Hajeb, P.; Zhu, L.; Bossi, R.; Vorkamp, K. Sample preparation techniques for suspect and non-target screening of emerging contaminants. *Chemosphere* **2022**, *287*, 132306. [CrossRef] [PubMed]
24. Lopez, S.H.; Dias, J.; Mol, H.; de Kok, A. Selective multiresidue determination of highly polar anionic pesticides in plant-based milk, wine and beer using hydrophilic interaction liquid chromatography combined with tandem mass spectrometry. *J. Chromatogr. A* **2020**, *1625*, 461226. [CrossRef] [PubMed]
25. Tan, S.; Yu, H.; He, Y.; Wang, M.; Liu, G.; Hong, S.; She, Y. A dummy molecularly imprinted solid-phase extraction coupled with liquid chromatography-tandem mass spectrometry for selective determination of four pyridine carboxylic acid herbicides in milk. *J. Chromatogr. B* **2019**, *1108*, 65–72. [CrossRef] [PubMed]
26. Samsidar, A.; Siddiquee, S.; Shaarani, S.M. A review of extraction, analytical and advanced methods for determination of pesticides in environment and foodstuffs. *Trends Food Sci. Technol.* **2018**, *71*, 188–201. [CrossRef]
27. Perestrelo, R.; Silva, P.; Porto-Figueira, P.; Pereira, J.A.; Silva, C.; Medina, S.; Câmara, J.S. QuEChERS-Fundamentals, relevant improvements, applications and future trends. *Anal. Chim. Acta* **2019**, *1070*, 1–28. [CrossRef]
28. *SANTE/12682/2019*; Analytical Quality Control and Method Validation Procedures for Pesticides Residues and Analysis in Food and Feed. Directorate General for Health and Food Safety. European Union: Brussels, Belgium, 2020.
29. González-Curbelo, M.Á.; Socas-Rodríguez, B.; Herrera-Herrera, A.V.; González-Sálamo, J.; Hernández-Borges, J.; Rodriguez-Delgado, M.A. Evolution and applications of the QuEChERS method. *TrAC Trends Anal. Chem.* **2015**, *71*, 169–185. [CrossRef]
30. Anagnostopoulos, C.; Bourmpopoulou, A.; Miliadis, G. Development and validation of a dispersive solid phase extraction liquid chromatography mass spectrometry method with electrospray ionization for the determination of multiclass pesticides and metabolites in meat and milk. *Anal. Lett.* **2013**, *46*, 2526–2541. [CrossRef]

Article

Analysis of Multiclass Pesticide Residues in Tobacco by Gas Chromatography Quadrupole Time-of-Flight Mass Spectrometry Combined with Mini Solid-Phase Extraction

Rui Bie [1,†], Jiguang Zhang [1,†], Yunbai Wang [1], Dongmei Jin [2], Rui Yin [3], Bin Jiang [4] and Jianmin Cao [1,*]

1. Laboratory of Quality & Safety Risk Assessment for Tobacco, Ministry of Agriculture and Rural Affairs, Tobacco Research Institute of Chinese Academy of Agricultural Sciences, Qingdao 266101, China; 82101202172@caas.cn (R.B.); zhangjiguang@caas.cn (J.Z.); wangyunbai@caas.cn (Y.W.)
2. Sichuan Tobacco Quality Supervision and Testing Station, Chengdu 610041, China; dongmeijin1979@outlook.com
3. Ningqiang Branch of Hanzhong Tobacco Company, China National Tobacco Corporation, Ningqiang 724400, China; yinruibsy@outlook.com
4. Shandong Branch of China National Tobacco Corporation, Jinan 250101, China; jiangbin8077@126.com
* Correspondence: caojianmin@caas.cn
† These authors contributed equally to this work.

Abstract: A screening method using gas chromatography quadrupole time-of-flight mass spectrometry (GC-QTOF/MS) combined with mini solid-phase extraction (mini-SPE) was established for the quantification and validation of multiclass pesticide residues in tobacco. The method was quicker and easier, with sample purity higher than that obtained by traditional SPE and dispersed-SPE. Box-Behnken design, an experimental design for response-surface methodology, was used to optimize the variables affecting the target pesticide recovery. Under the optimized conditions, 92% of the pesticides showed satisfactory recoveries of 70%–120% with precision <20% at spiking levels of 50, 250, and 500 ng/g. The limits of detection and quantification for all the analyses were 0.05–29.9 ng/g and 0.20–98.8 ng/g, respectively. In addition, a screening method based on the retention time and a homebuilt high-resolution mass spectrometry database were established. Under the proposed screening parameters and at spiking levels of 50, 100, and 500 ng/g, 76.6%, 94.7%, and 99.0% multiclass pesticide residues were detected, respectively, using the workflow software. The validated method was successfully applied to the analysis of real tobacco samples. Thus, the combination of mini-SPE and GC-QTOF/MS serves as a suitable method for the quantitative analysis and rapid screening of multiclass pesticide residues in tobacco.

Keywords: mini solid-phase extraction; multiclass pesticide residues; tobacco; gas chromatography quadrupole time-of-flight mass spectrometry

Citation: Bie, R.; Zhang, J.; Wang, Y.; Jin, D.; Yin, R.; Jiang, B.; Cao, J. Analysis of Multiclass Pesticide Residues in Tobacco by Gas Chromatography Quadrupole Time-of-Flight Mass Spectrometry Combined with Mini Solid-Phase Extraction. *Separations* 2022, *9*, 104. https://doi.org/10.3390/separations9050104

Academic Editor: Beatriz Albero

Received: 31 March 2022
Accepted: 16 April 2022
Published: 21 April 2022

Publisher's Note: MDPI stays neutral with regard to jurisdictional claims in published maps and institutional affiliations.

Copyright: © 2022 by the authors. Licensee MDPI, Basel, Switzerland. This article is an open access article distributed under the terms and conditions of the Creative Commons Attribution (CC BY) license (https://creativecommons.org/licenses/by/4.0/).

1. Introduction

Tobacco is a non-food crop, and its production heavily relies on the use of pesticides (including insecticides, herbicides, fungicides, and suckercides). Pesticide residues are the pesticides remaining on tobacco after harvesting. Studies have revealed that pesticides are present in the cigarette smoke, thus exposing both active and passive smokers to pyrolyzed pesticide residues [1,2]. The detection and removal of pesticide residues in tobacco have always been challenging, various countries and international organizations have established maximum residue limits for these residues in tobacco. For example, in 2021, the CORESTA Agro-Chemical Advisory Committee provided guidance residue levels (GRLs) for 117 pesticides and other chemicals in tobacco [3].

Various studies have reported the analysis of multiclass pesticide residues in tobacco by gas chromatography tandem mass spectrometry [4–6] and liquid chromatography tandem mass spectrometry [7–9]. In particular, gas chromatography quadrupole time-of-flight mass spectrometry (GC-QTOF/MS) has become an effective tool for the quantitative and high-throughput screening of targeted and non-targeted trace-level compounds in complex matrix samples. Thus, this technique has been employed by various researchers for the screening and quantification of pesticide residues in various food matrices [10–13]. However, only a few studies have reported the screening and quantification of multiclass pesticides in tobacco.

Because tobacco is a complex matrix and has high contents of pigments, terpenes, alkaloids, and flavonoids [4], a pretreatment step is required before detection to improve the sensitivity and specificity of the detection method. The pretreatment techniques that have been mainly used in the past decades are QuEChERS (Quick Easy Cheap Effective Rugged Safe) and SPE (Solid Phase Extraction) [4,14–17]. The QuEChERS method is quick, easy, inexpensive, effective, robust, and safe and can be used for the analysis of a large number of samples. However, this method has limited ability to eliminate matrix interferences, thus resulting in contamination. In contrast, SPE has remarkable cleanup efficiency, with higher accuracy and precision; however, this method requires multiple steps, making it a time-consuming, complex, and relatively expensive pretreatment method. Simple extraction methods have been gaining attention recently, among which mini-SPE is more effective owing to its simplicity, high extraction rate, and low consumption of organic solvents. This method has already been used as a pretreatment technique in the analysis of multi-pesticide residues in complex food matrices and spices [18,19]. However, mini-SPE has not yet been applied for the pretreatment of pesticide residues in tobacco. In this study, several important parameters affecting the performance of mini-SPE were optimized. In addition, the chromatograms of the tobacco extract cleaned up by mini-SPE and QuEChERS were compared to determine the cleanup efficiency of mini-SPE. Finally, a method for the screening and quantification of 209 pesticides in tobacco was developed using GC-QTOF/MS coupled with mini-SPE.

2. Material and Methods

2.1. Reagents and Materials

HPLC-grade ethyl acetate (cas: 141-78-6), acetic acid (cas: 64-19-7), acetonitrile (cas: 75-05-8), acetone (cas: 67-64-1), and *n*-hexane (cas: 110-54-3) were purchased from AN-PEL Laboratory Technologies Inc. (Shanghai, China). All pesticide reference standards (purity ≥ 95%) were obtained from Dr. Ehrenstorfer GmbH (Dr. Ehrenstorfer GmbH, Augsburg, Germany) and the Agro-Environmental Protection Institute, Ministry of Agriculture (Tianjin, China). The CAS numbers of all the pesticides are detailed in Figure S1 in Supplementary Materials. The mini-SPE column was purchased from Agela Technologies, Inc. (Tianjin, China). A schematic diagram of the mini-SPE apparatus is shown in Figure 1.

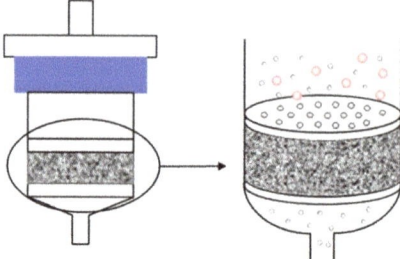

Figure 1. Schematic diagram of the mini-SPE.

2.2. Standard Solution Preparation

Primary stock solutions of each pesticide (1000 μg/mL) and the Mirex internal standard solution (5 μg/mL) were prepared in an n-hexane–acetone mixture (1:1, v/v). Based on the chemical properties and retention times of each pesticide, the 209 pesticides were divided into four groups: I, II, III, and IV. Stock solutions of mixed pesticide standards (1 μg/mL) were also prepared in the same n-hexane–acetone mixture (1:1, v/v). The matrix-matched standards (0.01, 0.05, 0.1, 0.2, 0.5, and 0.8 μg/mL) were prepared by diluting the mixed standards of each analyte with a blank matrix extract solution and a Mirex internal standard solution. All solutions were stored at $-20\ °C$ in a refrigerator.

2.3. Sample Preparation

A tobacco sample (1 g) was weighed into a centrifuge tube (50 mL), to which a Mirex internal standard solution (100 μL) and an acetonitrile-0.1% acetic acid solution (10 mL) were added. The centrifuge tube was vortexed at 2000 rpm for 10 min, and then centrifuged at 4000 rpm for 10 min. A 2 mL syringe was used for the mini-SPE column. The supernatant (1 mL) was loaded in the syringe and then slowly released. The effluent was then collected in a 2 mL centrifuge tube, concentrated using a vacuum concentrator at 45 °C, and finally reconstituted in n-hexane/acetone mixture (0.5 mL; 1:1, v/v). After vortexing for 30 s, the reconstituted solution was filtered through a 0.22-μm Nylon membrane prior to GC-QTOF/MS.

To verify the purification efficiency of mini-SPE, the National Standard of the People's Republic of China for the determination of pesticides and metabolites in foods of plant origin, published in 2018, was used for tobacco sample pretreatment. The extraction and cleanup procedures are described in detail in Figure S2 in Supplementary Materials.

2.4. GC-QTOF/MS

In this study, an Agilent 7890B GC system coupled to an Agilent 7200 Q-TOF mass spectrometer (Santa Clara, CA, USA) was used. Two HP-5MS capillary columns (15 m × 250 μm × 0.25 μm; Santa Clara, CA, USA) were connected by a backflush system, which was used at 40.5 min under 50 psi. The oven temperature program was as follows: initial temperature of 60 °C (1 min), increased to 120 °C at 40 °C/min, then to 310 °C at 5 °C/min, and then held for 5 min at 310 °C. The injection volume was 1.0 μL in the splitless mode. The inlet temperature was set to 250 °C. Helium (purity: 99.999%) was used as the carrier gas at a flow rate of 1 mL/min. To correct the retention time drift caused by the change in chromatographic column efficiency, Mirex was used for retention time locking.

Q-TOF/MS was operated in the EI mode with an electron energy of 70 eV. The high-resolution mode of 4 GHz (12000 FWHM), at which the TOF-MS system operates in the full-scan mode (m/z 50–500) at a rate of 5 spectra/s, allows more accurate analyte identification. Internal mass calibration with perfluorotributylamine (PFTBA) was performed before each injection to achieve a precise high-resolution and accurate mass operation. The temperatures of the transfer line, quadrupole, and ion source were maintained at 280 °C, 180 °C, and 230 °C, respectively. The analysis was performed with a solvent delay of 4 min to prevent damage to the filament. Data analysis was performed using Agilent MassHunter Version B.07.06. A mass spectrometry database was created using the Personal Compound Database and Library (PCDL) Manager (Version B.07.00, Agilent, Santa Clara, CA, USA). MassHunter Qualitative Analysis Workflow software (version B.08.00, Agilent, Santa Clara, CA, USA) was used to screen non-targeted pesticides based on a created accurate-mass spectrometry library. Agilent MassHunter quantitative analysis version (version B.09.00, Agilent, Santa Clara, CA, USA) was used for the quantitative determination of the targeted pesticides.

2.5. Experimental Design

Based on our previous single-factor experimental results, three important factors (water volume (A), solvent volume (B), and purification volume (C)) affecting the target pesticide recovery were studied using the Box-Behnken test design with the Design-Expert software (Table 1, version 13, Stat-Ease, Minneapolis, MN, USA).

Table 1. Factors and codes of the sample preparation procedure by the Box-Behnken design.

Factor	Code	Coding Level		
		−1	0	1
Water volume (mL)	A	0	2	4
Solvent volume (mL)	B	5	10	15
Purification volume (mL)	C	0.8	1.2	1.6

2.6. Analytical Parameters

The proposed method was validated in terms of recovery, linearity, limit of detection (LOD), limit of quantification (LOQ), and precision (coefficient of variation (CV)). The linearity of the method was determined using the matrix-matched standards (10, 50, 100, 200, 500, and 800 ng/mL). Residue-free tobacco samples were added to 50, 250, and 500 ng/g mixed standard stock solutions using three replicates to calculate the average recovery and CV of each pesticide. The LODs and LOQs of the method were calculated at signal-to-noise ratios (S/N) of 3 and 10, respectively.

3. Results and Discussion

3.1. Optimization of Extraction Conditions

The pesticides investigated in this study include organochlorines, organophosphorus, and carbamates, which have large differences in solubility and polarity. Therefore, a solvent with high solubility is needed for a more efficient extraction of the pesticides. To achieve high-efficiency extraction, the amount of matrix compounds co-extracted from the complex tobacco matrix should be as low as possible. According to literature, *n*-hexane–acetone mixture, ethyl acetate, acetonitrile, and acetonitrile–0.1% acetic acid are the most commonly used extraction solvents [20–23]. In this study, the effects of these four solvents on the 209 pesticide residues in tobacco were investigated; the recovery ranges of the target pesticides obtained upon extraction by these solvents are shown in Figure 2. The results showed that when acetonitrile–0.1% acetic acid was used as the extraction solvent, the proportion of pesticides with recoveries in the range of 60–120% was the largest. Moreover, as shown in Figure 3, the use of acetonitrile–0.1% acetic acid as the extraction solvent resulted in significant improvement in the recovery of some carbamates and organophosphorus pesticides with strong polarity, such as mevinphos, disulfoton, and methiocarb and a drift in the retention time was observed. This result was in accordance with that of a previous study [24]. Thus, acetonitrile–0.1% acetic acid was chosen as the optimal extraction solvent.

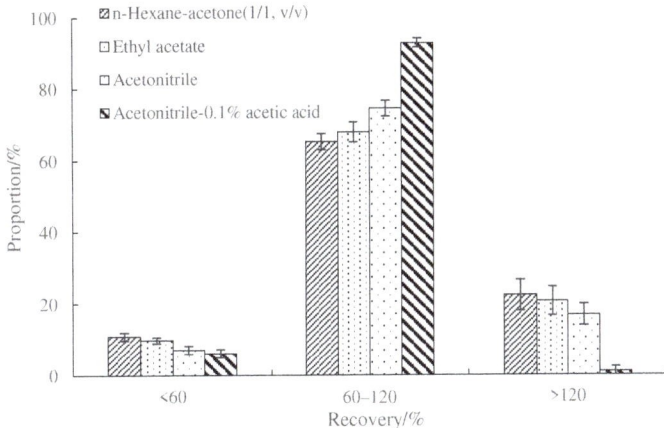

Figure 2. Recovery ranges of pesticides extracted by different solvents.

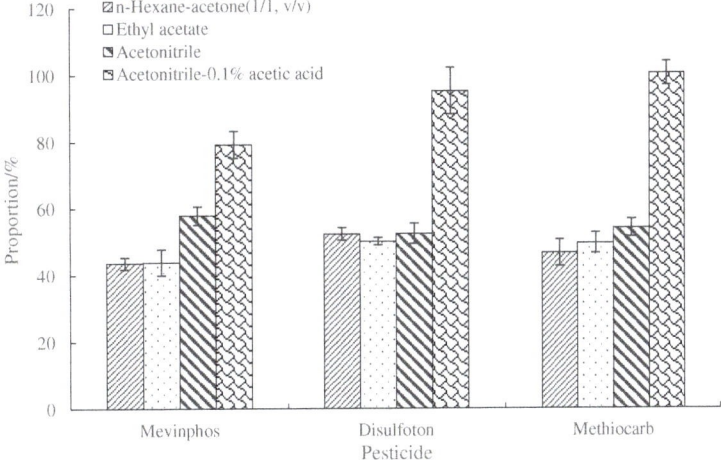

Figure 3. Recovery of three pesticides extracted by different solvents.

3.2. Optimization of Sample Preparation Conditions

According to the analysis of the response surface experimental data using the Design-Expert software, the regression equation indicating the relation of the proportion (Y) of the target pesticide, with a recovery in the range of 60%–120%, with various factors can be expressed by Equation (1):

$$Y = 77.14 - 15.82A - 2.93B - 3.19C - 2.3AB + 1.28AC - 3.06BC + 0.71A^2 - 9.75B^2 - 8.21C^2 \quad (1)$$

The results of ANOVA analysis indicate that the model was extremely significant ($p = 0.0012 < 0.01$), with an insignificant lack of fit ($p = 0.0981 > 0.05$), indicating that the regression equation and actual fitting had a small proportion of abnormal errors. The regression coefficient (R^2) value (0.9451) indicated a good model correlation. The coefficient of variation (7.15%) indicated high experimental stability. Within the selected range of factors, the p-values of A, B^2, and C^2 were <0.05, indicating that all factors had a significant impact on the pesticide recovery. Three-dimensional response surface plots of the predicted mode are shown in Figures 4–6.

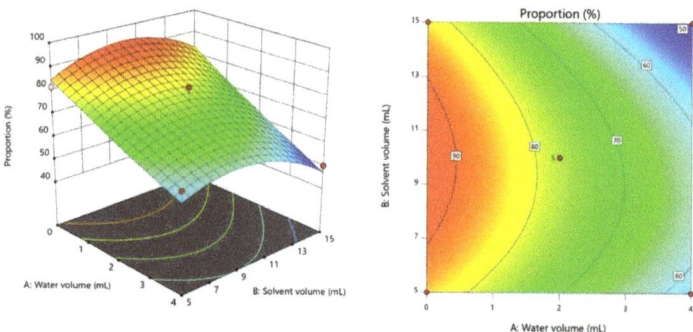

Figure 4. Response surface and contour diagram of R = f (A, B) with a 1.2 mL purification volume.

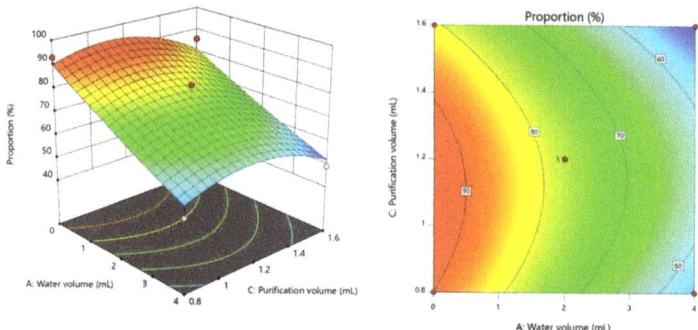

Figure 5. Response surface and contour diagram of R = f (A, C) with a 10 mL solvent volume.

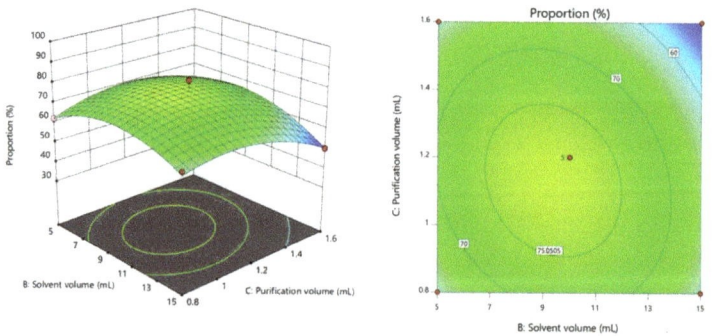

Figure 6. Response surface and contour diagram of R = f (B, C) with a 2 mL water volume.

3.3. Matrix Effects

The nature of the pesticide matrix affects the accuracy and repeatability of the results of GC-MS, and most pesticides exhibit different levels of matrix enhancement. In fact, during sample detection, impurities in the sample can compete with pesticide molecules for the active sites in the mass spectrometer inlet and column head, resulting in an increase of the target molecules. Therefore, the response of analytes with the same content in the matrix solution becomes higher than that in the pure solvent [25,26]. The matrix effect is closely related to the chemical structure and properties of analytes. Generally, the thermal instability, polarity, and hydrogen bonding ability of pesticides have a strong matrix effect in GC. In our previous work, the peak areas of the target pesticides in pure solvents and matrix solutions were compared at the same concentration, and most pesticides were found to exhibit matrix enhancement effects [5]. Therefore, matrix-matched calibration curves were chosen to nullify the matrix effect.

3.4. Screening Method

Under optimized chromatographic and mass spectrometric conditions, high-resolution mass spectrograms of the 209 pesticides were collected in the full-scan mode and imported to the PCDL software. The name, retention time, molecular formula, accurate mass, CAS number, and structural formula were imported to the PCDL software to establish an accurate mass spectrometry library. In the Agilent MassHunter Qualitative Analysis Workflow software, the homebuilt library was selected, and the search parameters were set as follows: retention time deviation, ±0.15 min; accurate mass deviation, ±20 ppm; minimum qualified fragment number, 2; co-elution matching score, s ±20 ppm. The minimum number of qualified fragment ions measured for each compound and the theoretical value in the library based on the retention time, accurate mass deviation, isotope peak distribution, and abundance ratio were calculated, and a matching score was assigned. Further, the ratio of qualitative and quantitative ions is an important parameter to judge the existence of false positives. The default matching score of qualitative and quantitative ion ratio in the workflow software was ≥75. Therefore, although the co-elution matching score was ≥70, the matching score of qualitative and quantitative ion ratio was <75. The software will provide a warning, indicating the possibility of a false positive. At this point, manual verification is necessary.

The screening method was validated using blank samples spiked with 50, 100, and 500 ng/g pesticides. The screening was performed as described above, and the proportion of pesticides screened at the three concentrations was calculated. The results showed that the proportion of pesticides detected by the workflow software was 76.6%, 94.7%, and 99.0% at the three concentrations, respectively, under the proposed screening parameters. The screening limit of this method was higher than that reported in other studies [27,28], mainly because of the difference between the evaluation method and the pretreatment process. In these studies, a blank matrix matching a mixed standard solution was used for direct injection when the screening limit was evaluated. In this work, the blank samples were spiked with different concentrations of pesticide-mixed standard solutions, and then extracted and purified using the above-mentioned method. This was in accordance with the test requirements for real samples. Owing to the large dilution ratio in the pretreatment process, the final sample concentration detected in the test solution was 0.2 g/mL. During sample analysis, the screening ability of the method can be improved by increasing the concentration ratio.

3.5. Comparison of Cleanup Efficiency of Mini-SPE and d-SPE

The total ion chromatography (TIC) chromatograms of the tobacco extracts obtained by mini-SPE (black) and dispersed-SPE (d-SPE; red) are shown in Figure 7. The mini-SPE-treated sample showed a lower TIC chromatographic baseline, indicating the stronger ability of mini-SPE to remove impurity interferences, especially alkaloids such as nicotine, nicotyrine, and (R,S)-anatabine. In addition, the ability of mini-SPE to remove megastigmatrienone-I, II, III, and IV, which are important aroma components in tobacco, was stronger, with only a small amount of megastigmatrienones present in the solution after extraction. The two purification methods showed similar abilities for the removal of 4,8,13-duvatriene-1,3-diol, a major glandular trichome secreted by tobacco. In general, mini-SPE can effectively clean up alkaloids, aroma components, and pigments. Moreover, mini-SPE requires few steps and is simple, making it an excellent pretreatment method for rapid detection.

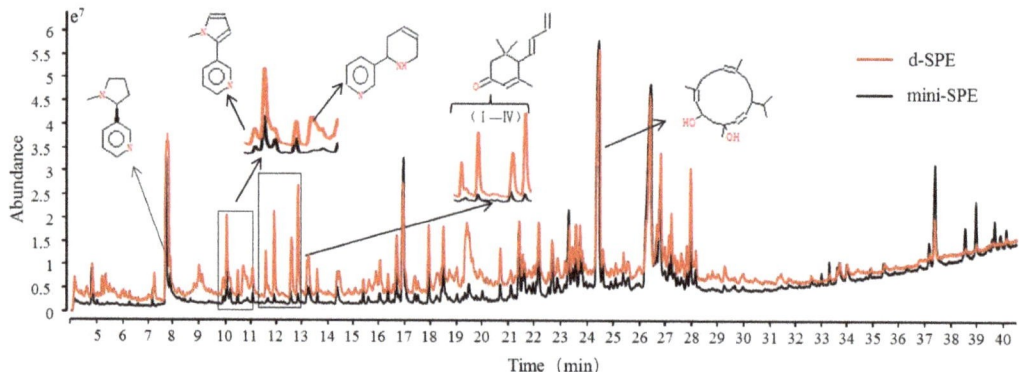

Figure 7. Total ion current (TIC) chromatogram of blank tobacco extract extracted with mini-SPE (black) and with d-SPE (red) method.

3.6. Analytical Parameters of Quantitative Method

The linear regression coefficients (r^2), LOD, LOQ, recovery, and CV values of the 209 pesticides are listed in Table A1 in the Appendix A. The r^2 were higher than 0.995 in the linearity range of 10–800 ng/mL for all the 209 pesticides tested. The LODs for all analyses ranged from 0.05 to 29.8 ng/g, while the LOQs were in the range of 0.2–98.9 ng/g. At spiking levels of 50, 250, and 500 ng/g, the recoveries of the pesticides were 64.2%–122.1%, 66.8%–124.0%, and 63.8%–127.7%, respectively, except for 1,2-dibromo-3-chloropropane, naled, and hexachlorobenzene; the CVs were 0.3%–15.5%, 0.1%–14.3%, and 0.28%–11.9%, respectively. 1,2-Dibromo-3-chloropropane and naled are very unstable and are easily decomposed upon heating [29,30], while other pesticides can decompose in the GC system during the injection process, resulting in a lower recovery of the two pesticides. The recovery of hexachlorobenzene was also very low, which can be attributed to its planar structure, similar to that of the purification material.

3.7. Real Sample Analysis

To demonstrate the applicability of the developed and validated method for the analysis of real samples, seven tobacco samples obtained from three main planting regions in China were analyzed for their pesticide residues. Tobacco samples were prepared according to the method described in Section 2.3. sample preparations, and determined in full scan mode by GC-Q-TOF. Then the screening method was applied to screen the pesticides. There were 12 output results with a screening score greater than 70, involving 7 pesticides, and the qualitative and quantitative ion ratios of all output results were greater than 75. Only 2 of the 7 samples did not detect pesticide residues. Seven pesticides were metalaxyl, triadimefon, triadimenol, dimetachlone, myclobutanil, flumetralin, and cyhalothrin, which are classified as fungicides and insecticides. The detected pesticides were quantified using matrix-matched calibration standards, all of which were below the GRLs set by CORESTA. The results obtained by the proposed method were compared with those obtained by the GC-MS/MS method used in our laboratory [5]. The CV values of the quantitative results obtained using the two methods ranged between 5% and 13.35%.

4. Conclusions

In this study, a simple and rapid sample-preparation method coupled with GC-Q/TOF technique was developed for the screening and quantification of 209 pesticides in tobacco. When this method was evaluated on 209 pesticides in tobacco, 192 of them showed satisfactory recovery and precision at the spiked levels of 50, 250, and 500 ng/g. In the process of sample pretreatment, mini-SPE technology was used to purify tobacco samples for the first time. Compared with the traditional SPE, the samples in mini-SPE can be loaded and eluted directly, without any activation/equilibration, cleaning, or elution step, which greatly reduced the sample pretreatment time and the amount of organic solvent. Mini-SPE requires few steps, making it an excellent pretreatment method for rapid detection. Moreover, mini-SPE also exhibits good cleanup efficiency, the comparative test showed that mini SPE had stronger ability to remove the pigments and alkaloids than d-SPE. Furthermore, this method was found to be applicable for the analysis of real samples, demonstrating its suitability for sensitive and rapid screening of pesticide residues. The developed method provided accurate and reliable quantitative screening results, was simple and fast, and could be used for the analysis of multiclass pesticide residues in tobacco.

Supplementary Materials: The following supporting information can be downloaded at: https://www.mdpi.com/article/10.3390/separations9050104/s1, Figure S1: CAS number, retention time of the 209 pesticides; Figure S2: Sample preparation of d-SPE.

Author Contributions: Conceptualization, Y.W. and J.C.; methodology, R.B. and J.C.; validation, J.Z. and R.Y.; formal analysis, J.C. and J.Z.; investigation, D.J.; data curation, B.J. and R.Y.; writing—original draft preparation, R.B. and J.Z.; writing—review and editing, R.B., J.Z. and J.C. All authors have read and agreed to the published version of the manuscript.

Funding: This work was funded by the Agricultural Science and Technology Innovation Program (ASTIP-TRIC06), the National Agricultural Product Quality and Safety Risk Assessment Project (GJFP2019018).

Conflicts of Interest: The authors declare no conflict of interest.

Appendix A

Table A1. The r^2, LOD, LOQ, recovery, and CV values of the 209 pesticides.

Pesticides	r^2	LOD (ng/g)	LOQ (ng/g)	Spiked at 50 ng/g		Spiked at 250 ng/g		Spiked at 500 ng/g	
				Recovery (%)	CV (%)	Recovery (%)	CV (%)	Recovery (%)	CV (%)
1,2-Dibromo-3-chloropropane	0.997	4.5	14.8	40.3	3.6	49.6	5.7	49.3	9.6
Dichlorvos	0.997	4.4	14.6	100.4	7.1	72.7	3.7	80.4	11.8
Disulfoton sulfoxide	0.999	1.4	4.5	96.2	7.0	106.2	1.2	104.3	3.8
Mevinphos(E)	0.999	4.9	16.1	83.3	8.9	75.3	3.0	78.9	9.2
Butylate	0.998	2.2	7.2	71.4	5.4	107.0	7.0	117.1	6.5
Mevinphos(Z)	0.998	6.2	20.6	109.2	2.8	74.9	2.9	77.3	7.6
Pebulate	0.995	8.3	27.2	96.8	10.2	109.8	2.2	111.1	2.1
Methacrifos	0.997	3.3	11.0	110.2	8.9	114.0	3.6	106.3	2.1
Molinate	0.998	3.8	12.7	90.5	3.0	82.0	1.6	107.2	5.1
Isoprocarb	0.998	10.6	35.1	83.0	6.4	80.0	6.3	100.2	8.9
Heptenophos	0.998	3.2	10.4	100.7	10.1	110.1	2.8	120.7	7.6
Chlorphenprop-methyl	0.997	7.1	23.4	91.2	15.5	82.4	6.1	88.6	6.6
Thionazin	0.997	8.1	26.7	95.9	10.6	110.8	9.4	110.9	7.4
Fenobucarb	0.998	4.4	14.5	79.9	11.5	78.6	4.4	101.4	6.8
Propoxur	0.998	3.0	9.9	91.5	12.5	81.8	4.4	113.4	0.3
Demeton-O	0.998	7.2	23.9	88.6	3.8	100.5	4.5	98.2	3.4
Demeton-S-methyl	0.999	9.7	31.9	77.9	9.5	108.1	7.0	89.1	3.8
Cycloate	0.997	2.7	8.9	116.2	3.1	111.9	11.3	108.7	1.3
Ethoprophos	0.999	4.6	15.3	104.5	7.8	119.8	0.7	122.3	3.2
Chlorpropham	0.996	7.5	24.8	70.7	0.3	88.1	12.3	86.4	9.4
Naled	0.996	7.1	23.3	43.8	12.5	50.2	6.4	52.4	5.3
Chlordimeform	0.999	9.5	31.4	73.4	1.5	78.1	6.5	75.3	4.0
Trifluralin	0.995	0.8	2.7	102.6	11.5	121.2	6.1	126.3	9.7
Benfluralin	0.995	1.3	4.3	82.6	6.5	95.8	4.8	111.5	4.5
Cadusafos	0.999	1.5	5.0	99.2	5.4	116.8	3.3	117.6	1.5
Phorate	0.995	0.5	1.6	107.3	8.0	106.6	10.6	106.5	7.7
BHC-alpha	0.998	2.0	6.5	91.4	7.8	115.5	5.0	109.5	7.8
Hexachlorobenzene	0.996	0.3	0.9	27.5	13.6	35.4	5.6	46.5	1.0
Dicloran	0.999	11.2	37.0	73.2	5.5	86.5	4.2	80.5	0.7
Demeton-S	0.999	10.4	34.2	87.4	8.0	88.3	2.1	113.5	7.7
Dimethoate	0.998	9.2	30.5	71.4	4.7	81.4	10.8	76.2	6.9
Carbofuran	0.996	8.4	27.8	99.9	9.9	101.1	5.5	105.5	5.3

Table A1. Cont.

Pesticides	r^2	LOD (ng/g)	LOQ (ng/g)	Spiked at 50 ng/g		Spiked at 250 ng/g		Spiked at 500 ng/g	
				Recovery (%)	CV (%)	Recovery (%)	CV (%)	Recovery (%)	CV (%)
Atrazine	0.996	1.7	5.6	87.4	10.8	83.4	2.8	91.2	5.7
BHC-beta	0.999	1.3	4.1	74.1	8.0	95.0	10.2	97.6	6.2
Clomazone	0.998	2.0	6.5	119.2	5.2	114.6	3.7	114.5	4.7
Propazine	0.998	11.6	38.3	94.4	2.4	83.9	9.1	93.9	3.6
Terbumeton	0.996	8.0	26.3	77.7	5.1	78.2	4.4	84.1	3.6
BHC-gamma	0.998	1.3	4.4	104.3	11.8	112.6	11.1	110.1	3.9
Quintozene	0.999	0.1	0.3	71.6	8.1	110.2	7.2	98.0	3.9
Terbufos	0.997	4.5	14.7	99.3	3.4	107.2	2.2	124.8	4.4
Trietazine	0.998	5.5	18.3	73.3	3.2	76.9	3.9	85.7	2.1
Fonofos	0.998	6.7	22.1	110.1	5.6	112.4	2.4	116.8	0.6
Phosphamidon(E)	0.998	11.8	38.8	85.9	6.3	113.2	3.3	97.3	3.3
Diazinon	0.999	5.8	19.1	116.6	9.9	107.1	3.3	110.6	4.0
Disulfoton	0.999	7.5	24.9	102.5	5.5	88.5	6.4	94.6	6.7
BHC-delta	0.995	13.7	45.2	111.1	7.8	103.3	12.1	112.0	5.1
Mexacarbate	0.996	4.3	14.3	94.4	5.5	96.2	4.9	114.2	8.9
Triallate	0.997	1.8	5.8	88.4	5.8	117.1	3.4	116.2	4.1
Tefluthrin	0.996	3.1	10.1	116.5	1.2	116.9	1.5	127.7	3.0
Isazofos	0.997	3.1	10.3	103.3	12.4	115.2	1.6	119.8	4.6
3-Hydroxycarbofuran	0.995	7.7	25.3	108.2	4.9	107.8	1.8	118.9	7.7
Iprobenfos	0.998	3.2	10.6	104.2	5.1	121.3	3.3	124.7	6.0
Pirimicarb	0.998	2.1	6.9	93.1	2.3	108.3	2.6	116.5	1.4
Benfuresate	0.998	4.0	13.3	90.5	8.4	79.5	8.8	85.7	5.1
Phosphamidon(Z)	0.999	6.4	21.2	101.9	3.7	78.3	3.3	76.7	1.3
Propanil	0.997	7.9	26.2	86.6	5.6	76.1	7.6	87.4	8.1
Dimethachlor	0.996	9.2	30.4	76.5	8.1	78.8	6.4	84.9	5.1
Acetochlor	0.998	19.0	62.8	93.8	7.4	110.1	4.9	99.6	2.6
Parathion-methyl	0.996	5.4	18.0	110.2	5.0	99.5	7.3	78.5	1.4
Chlorpyrifos-methyl	0.998	3.6	11.7	94.8	7.2	88.0	2.1	97.9	4.9
Vinclozolin	0.999	6.5	21.4	99.8	6.7	106.0	11.4	91.0	2.6
Simetryn	0.998	14.1	46.4	104.1	11.9	70.0	2.7	72.4	3.2
Carbaryl	0.996	14.5	47.7	73.2	3.5	77.2	2.8	107.4	9.8
Tolclofos-methyl	0.998	6.3	20.9	103.5	7.0	110.9	1.5	113.1	4.2
Heptachlor	0.999	0.4	1.4	104.7	4.7	111.0	7.5	102.2	1.6
Alachlor	0.999	4.3	14.3	100.2	1.0	97.7	8.8	104.5	3.7
Prometryn	0.999	8.4	27.7	121.7	9.3	106.0	4.4	103.1	5.4
Metalaxyl	0.996	3.9	12.8	105.8	8.1	93.7	6.4	116.0	7.3
Fenchlorphos	0.999	6.2	20.4	100.7	6.4	105.3	4.4	117.5	1.6
Prosulfocarb	0.997	5.6	18.3	91.6	6.1	72.3	0.1	104.8	5.1
Demeton-S-methylsulfone	0.996	14.5	47.9	76.4	8.0	85.7	0.8	87.7	1.6
Thiobencarb	0.998	6.7	22.2	98.1	7.6	86.2	9.4	100.4	3.3
Orbencarb	0.998	7.2	23.9	118.0	0.3	92.8	11.6	105.2	5.2
Methiocarb	0.996	2.9	9.5	95.8	7.1	98.1	7.4	104.6	8.8
Fenitrothion	0.998	11.0	36.4	113.7	11.3	113.9	12.8	110.7	3.6
Pentanochlor	0.998	8.0	26.4	100.7	3.7	74.5	2.3	101.3	4.1
Pirimiphos-methyl	0.999	14.5	47.8	108.2	12.0	114.4	3.0	113.2	3.8
Bromacil	0.999	2.3	7.6	88.4	11.8	110.9	12.2	105.1	6.1
Ethofumesate	0.997	8.3	27.5	85.5	1.4	86.0	6.3	88.6	5.3
Aldrin	0.999	6.5	21.6	113.2	1.0	110.8	1.8	107.1	4.4
Malathion	0.998	10.5	34.5	105.6	10.2	109.1	11.5	108.9	7.6
Phorate-sulfone	0.995	8.9	29.5	113.3	5.2	104.2	8.9	124.2	1.5
Metolachlor	0.999	5.4	17.8	107.6	4.2	112.1	1.5	111.4	4.7
Fenthion	0.999	3.9	12.7	99.4	8.6	108.7	8.7	114.8	2.7
Dicofol	0.997	20.3	67.1	94.8	3.1	103.5	0.3	113.1	3.9
Parathion	0.996	7.5	24.9	79.9	3.7	83.9	4.3	99.1	2.5
Thiazopyr	0.998	0.1	0.2	73.8	8.2	100.4	5.4	109.7	8.0
Chlorpyrifos	0.997	4.8	15.7	106.6	9.0	113.0	10.5	120.1	2.6
Triadimefon	0.996	9.6	31.7	105.7	7.9	122.3	2.9	120.6	6.0
Chlorthal-dimethyl	0.999	1.7	5.6	116.3	5.8	108.2	2.2	107.2	1.8
Flufenacet	0.998	9.8	32.3	106.4	9.9	111.1	2.2	108.3	7.5
Dimetachlone	0.999	5.0	16.4	117.7	7.7	82.1	3.8	103.6	5.8
Isocarbophos	0.998	4.4	14.4	118.5	1.1	104.3	6.0	105.1	8.3
Thiamethoxam	0.998	4.8	15.8	108.1	5.9	78.4	12.6	79.8	7.4
Bromophos	0.999	3.1	10.4	116.3	5.4	98.8	7.8	106.3	4.4
Butralin	0.996	2.9	9.6	109.0	5.7	96.7	10.4	103.8	4.2
Diphenamid	0.997	9.0	29.6	115.1	4.2	92.6	8.1	86.3	6.2
Isopropalin	0.995	8.1	26.7	70.0	5.6	78.4	9.1	71.0	4.9
Oxychlordane	0.999	1.8	5.8	103.0	11.5	88.3	5.9	90.9	4.2

Table A1. Cont.

Pesticides	r^2	LOD (ng/g)	LOQ (ng/g)	Spiked at 50 ng/g		Spiked at 250 ng/g		Spiked at 500 ng/g	
				Recovery (%)	CV (%)	Recovery (%)	CV (%)	Recovery (%)	CV (%)
trans-Chlorfenvinphos	0.996	2.3	7.7	87.3	11.3	105.0	3.6	99.4	7.9
Heptachlor epoxides (cis-)	0.997	3.6	11.7	106.9	8.4	111.8	11.1	98.7	5.2
Terbufos sulfone	0.997	2.6	8.4	91.9	4.6	114.5	7.0	123.3	1.8
Pendimethalin	0.995	9.6	31.6	105.7	3.4	123.0	3.5	118.1	1.8
Penconazole	0.999	10.0	33.1	114.5	2.2	101.8	4.7	101.1	3.5
Heptachlor epoxides (trans-)	0.998	5.5	18.3	110.8	1.4	98.5	3.1	82.3	6.5
Captan	0.998	10.9	36.1	80.3	6.6	74.5	5.4	86.9	2.0
cis-Chlorfenvinphos	0.998	5.6	18.5	114.3	4.1	117.4	3.3	119.7	0.4
Isofenphos	0.995	5.8	19.2	97.9	7.9	113.6	3.6	118.6	3.8
Quinalphos	0.999	3.3	11.0	88.4	9.3	102.4	3.8	106.4	6.2
Triadimenol	0.997	5.6	18.6	97.1	13.1	105.6	5.9	107.3	5.3
Phenthoate	0.996	7.4	24.3	81.8	8.1	101.7	14.0	109.8	2.4
Folpet	0.999	5.8	19.1	67.6	8.6	68.8	0.8	117.6	1.7
Methoprene	0.995	9.3	30.6	98.5	4.8	109.1	8.1	108.9	6.8
Chlordane-trans	0.998	2.0	6.5	83.5	7.0	115.7	5.9	116.4	2.6
Methidathion	0.995	8.3	27.4	109.8	13.0	124.0	6.5	119.6	1.1
o,p'-DDE	0.995	2.3	7.6	92.0	1.6	109.5	4.4	117.4	1.2
Haloxyfop-methyl	0.996	6.7	22.2	102.2	6.2	106.9	2.9	106.8	10.9
alpha-Endosulfan	0.997	4.6	15.0	83.3	6.1	119.5	1.7	109.6	3.9
Disulfoton-sulfone	0.997	9.3	30.8	118.3	6.5	110.7	10.1	115.2	2.3
Tetrachlorvinphos	0.998	4.3	14.3	90.7	12.5	109.1	9.2	103.6	8.2
Chlordane-cis	0.997	1.6	5.3	74.3	11.3	66.8	3.4	100.6	8.3
Mepanipyrim	0.999	7.9	25.9	68.0	6.5	75.5	0.4	103.9	8.8
Butachlor	0.999	10.2	33.5	109.0	6.5	108.8	5.8	115.1	6.9
Flumetralin	0.999	2.7	8.9	94.7	4.7	109.5	5.6	118.9	1.4
Napropamide	0.999	2.8	9.3	115.3	1.4	108.5	10.6	111.1	6.2
Fenamiphos	0.999	20.9	68.9	92.8	10.4	104.9	10.2	119.9	3.2
Butamifos	0.997	8.1	26.7	101.0	8.2	101.8	9.8	104.5	8.9
Hexaconazole	0.998	1.2	4.0	117.7	4.1	104.9	12.2	112.3	7.2
Imazalil	0.999	14.4	47.5	70.8	6.0	79.2	0.5	87.9	6.7
Prothiofos	0.999	3.5	11.6	102.4	2.1	96.7	8.8	123.3	2.1
Isoprothiolane	0.996	7.5	24.7	76.4	6.3	102.1	11.6	112.2	2.0
Profenofos	0.998	5.2	17.1	83.7	5.9	94.3	10.4	119.2	5.1
Dieldrin	0.998	5.9	19.5	74.8	9.0	100.2	8.0	88.4	8.1
p,p'-DDE	0.999	1.3	4.3	122.1	2.3	111.4	7.8	110.8	6.2
Uniconazole-P	0.997	0.4	1.4	104.8	10.8	93.2	6.6	88.1	6.2
Pretilachlor	0.999	0.8	2.6	98.1	2.3	108.4	0.5	105.5	6.1
Tribufos	0.999	7.4	24.4	116.9	8.5	110.6	10.1	114.1	3.9
Oxadiazon	0.997	9.3	30.5	92.8	7.6	94.2	6.5	101.5	3.3
o,p'-DDD	0.998	1.0	3.1	111.0	4.7	112.6	3.2	115.7	3.1
Myclobutanil	0.996	16.4	54.2	86.9	9.5	81.7	12.4	105.7	3.4
Flamprop-methyl	0.998	2.2	7.4	88.7	4.9	89.5	2.9	92.8	4.2
Buprofezin	0.995	11.3	37.2	80.3	5.4	94.4	6.5	101.5	8.4
Oxyfluorfen	0.998	5.0	16.5	76.4	5.0	102.4	9.9	112.4	4.0
Bupirimate	0.999	11.2	37.1	111.4	7.1	109.2	1.6	110.3	4.7
Thifluzamide	0.999	1.4	4.7	94.3	1.4	113.7	8.3	110.1	1.8
Kresoxim-methyl	0.997	6.9	22.7	82.8	9.1	99.3	7.3	99.5	3.8
Nitrofen	0.998	8.7	28.6	87.1	10.3	98.6	4.4	83.1	5.4
Endrin	0.999	4.8	15.9	90.6	6.2	97.5	8.9	103.3	0.5
Isoxathion	0.999	14.7	48.4	85.1	6.5	77.3	4.4	114.8	9.6
Fluazifop-butyl	0.997	8.0	26.6	67.1	2.7	94.7	10.4	73.2	1.8
beta-Endosulfan	0.999	18.1	59.6	115.9	2.8	96.4	3.5	63.8	6.7
Chlorobenzilate	0.999	7.4	24.4	107.9	10.2	107.5	7.4	106.8	9.2
Fensulfothion	0.999	2.4	7.9	108.3	8.1	77.6	6.6	104.1	2.6
Fenthion sulfoxide	0.998	13.6	44.7	96.3	10.3	107.1	4.6	111.5	2.5
Aclonifen	0.997	1.2	4.0	86.6	9.4	111.3	8.9	99.4	7.1
p,p'-DDD	0.999	2.4	8.0	105.0	2.5	99.9	2.3	104.9	8.2
Fenthion sulfone	0.998	12.8	42.2	86.3	11.4	117.3	10.2	96.3	9.4
o,p'-DDT	0.999	1.0	3.2	104.9	8.0	114.7	3.3	100.1	4.8
Oxadixyl	0.999	12.5	41.2	71.3	0.9	68.3	3.7	70.0	0.4
Ethion	0.997	5.2	17.2	120.8	3.7	119.3	1.1	116.2	4.6
Chlorthiophos	0.996	4.9	16.2	93.4	4.6	108.0	1.2	113.0	5.9
Triazophos	0.997	21.3	70.2	89.4	1.5	99.8	9.8	120.0	4.2
Carbophenothion	0.997	2.8	9.1	94.0	1.1	110.5	11.6	114.8	3.4
Benalaxyl	0.999	6.8	22.3	95.5	4.9	102.1	6.0	106.1	5.6
Endosulfan sulfate	0.998	1.1	3.7	98.3	5.6	106.0	2.1	105.1	3.9

Table A1. Cont.

Pesticides	r^2	LOD (ng/g)	LOQ (ng/g)	Spiked at 50 ng/g		Spiked at 250 ng/g		Spiked at 500 ng/g	
				Recovery (%)	CV (%)	Recovery (%)	CV (%)	Recovery (%)	CV (%)
Carfentrazone-ethyl	0.996	9.3	30.6	95.8	8.7	85.4	7.4	95.4	5.4
Propiconazole I	0.999	11.1	36.7	104.9	5.3	90.6	5.0	94.7	11.9
Propiconazole II	0.998	10.3	34.1	97.8	4.5	108.1	7.7	112.9	2.9
p,p'-DDT	0.999	2.8	9.1	81.6	8.2	89.1	8.5	80.7	5.7
Hexazinone	0.996	13.1	43.1	85.4	6.7	73.5	6.6	72.8	8.7
Tebuconazole	0.997	20.2	66.7	92.0	8.8	82.5	9.2	83.6	6.9
Thenylchlor	0.998	6.4	21.1	77.6	1.0	81.7	8.4	70.1	0.5
Triphenyl phosphate	0.999	4.9	16.0	105.0	6.7	81.2	3.8	79.8	6.1
Piperonyl butoxide	0.996	4.8	15.8	117.5	1.8	108.4	2.5	107.7	4.0
Pyributicarb	0.998	8.2	27.0	90.0	0.6	91.1	5.3	98.2	1.8
Benzoylprop-ethyl	0.996	2.4	7.8	86.7	2.0	91.5	6.0	90.9	2.9
Iprodione	0.997	11.7	38.8	74.8	10.3	74.0	0.4	106.3	7.4
Bromopropylate	0.998	3.5	11.7	95.6	8.7	93.1	6.9	103.4	2.7
Carbosulfan	0.998	1.8	5.9	95.9	6.6	108.4	2.2	87.9	10.9
EPN	0.999	13.7	45.1	96.1	11.1	112.9	5.0	114.3	5.9
Picolinafen	0.998	8.9	29.2	71.2	8.9	78.2	2.0	74.3	5.3
Chlorantraniliprole	0.996	10.9	35.9	73.8	3.1	72.3	6.2	73.3	5.2
Bifenthrin	0.997	8.4	27.7	94.8	4.3	104.5	3.4	120.5	3.2
Methoxychlor	0.999	5.0	16.6	86.2	8.0	104.6	3.6	110.4	4.9
Fenamidone	0.999	9.9	32.8	82.7	4.9	101.7	7.2	102.2	6.0
Anilofos	0.996	8.9	29.2	64.2	2.3	96.4	2.1	110.4	4.7
Clomeprop	0.998	4.4	14.4	81.1	8.5	71.3	1.7	76.0	9.8
Tetradifon	0.999	27.6	91.2	83.7	4.9	95.9	4.5	83.2	5.2
Phosalone	0.998	9.6	31.8	86.9	0.4	110.9	5.1	109.2	6.7
Leptophos	0.999	9.5	31.3	72.4	3.3	74.9	7.4	112.8	2.0
Cyhalofop-butyl	0.999	16.7	55.0	98.8	2.2	96.5	6.1	113.4	5.9
Cyhalothrin	0.997	20.1	66.4	101.7	8.7	95.4	7.4	110.7	5.0
Fenarimol	0.997	13.9	46.0	93.0	6.8	96.7	9.0	98.8	3.7
Pyrazophos	0.996	13.9	46.0	114.8	6.8	70.4	6.6	100.7	5.3
Benfuracarb	0.998	13.1	43.2	115.5	1.9	109.3	2.6	110.8	4.0
Fenoxaprop-P-ethyl	0.998	2.5	8.3	103.1	10.8	70.6	3.5	77.3	6.3
Bitertanol	0.996	26.9	88.9	87.4	5.2	96.5	3.8	100.6	1.3
Permethrin-cis	0.996	7.4	24.3	111.1	6.6	112.8	5.1	119.1	5.1
Permethrin-trans	0.997	6.9	22.7	91.4	11.8	110.2	1.2	113.2	5.1
Boscalid	0.998	8.5	28.2	83.8	6.3	84.3	5.3	82.0	3.1
Quizalofop-p-ethyl	0.999	9.6	31.7	101.8	8.7	100.6	6.3	95.7	4.9
Quizalofop-ethyl	0.996	8.9	29.3	110.2	8.7	99.7	9.5	111.6	2.1
Flucythrinate I	0.999	13.6	44.9	89.0	10.1	91.6	10.5	92.0	3.9
Flucythrinate II	0.999	14.4	47.5	114.0	12.3	99.8	6.7	91.4	5.5
Fenvalerate	0.997	29.9	98.8	107.9	8.1	96.1	4.4	91.7	4.7
Deltamethrin	0.999	14.1	46.6	80.3	6.3	93.0	3.3	98.2	6.9
Indoxacarb	0.998	14.2	47.0	90.2	11.5	104.6	14.3	105.0	6.4
Dimethomorph(Z)	0.995	12.6	41.5	87.3	8.0	104.9	6.6	115.0	5.1
Dimethomorph(E)	0.996	14.3	47.1	103.3	7.3	113.8	8.4	98.2	6.9

References

1. Gaddamidi, V.; Zimmerman, W.T.; Ponte, M.; Ruzo, L. Pyrolysis of C-14-Chlorantraniliprole in Tobacco. *J. Agric. Food Chem.* **2011**, *59*, 9424–9432. [CrossRef]
2. Davila, E.L.; Houbraken, M.; De Rop, J.; Wumbei, A.; Du Laing, G.; Romero, O.R.; Spanoghe, P. Pesticides residues in tobacco smoke: Risk assessment study. *Environ. Monit. Assess.* **2020**, *192*, 1–15. [CrossRef]
3. CORESTA. Guide: No.1 Agrochemical Guidance Residue Levels (GRLs). 2021, Agro-Chemical Advisory Committee of CORESTA. Available online: https://www.coresta.org/agrochemical-guidance-residue-levels-grls-29205.html (accessed on 25 December 2021).
4. Chen, X.S.; Bian, Z.Y.; Hou, H.W.; Yang, F.; Liu, S.S.; Tang, G.L.; Hu, Q.Y. Development and Validation of a Method for the Determination of 159 Pesticide Residues in Tobacco by Gas Chromatography-Tandem Mass Spectrometry. *J. Agric. Food Chem.* **2013**, *61*, 5746–5757. [CrossRef]
5. Khan, Z.S.; Girame, R.; Utture, S.C.; Ghosh, R.K.; Banerjee, K. Rapid and sensitive multiresidue analysis of pesticides in tobacco using low pressure and traditional gas chromatography tandem mass spectrometry. *J. Chromatogr. A* **2015**, *1418*, 228–232. [CrossRef]
6. Cao, J.M.; Sun, N.; Yu, W.S.; Pang, X.L.; Lin, Y.N.; Kong, F.Y.; Qiu, J. Multiresidue determination of 114 multiclass pesticides in flue-cured tobacco by solid-phase extraction coupled with gas chromatography and tandem mass spectrometry. *J. Sep. Sci.* **2016**, *39*, 4629–4636. [CrossRef]

7. Zhu, W.; Gao, C.; Lou, X.; Zhang, X.; Shi, X.; He, Y.; Wang, C. Rapid determination of 57 pesticide residues in tobacco leaves by LC-MS/MS. *Acta Tab. Sin.* **2013**, *19*, 12–16. [CrossRef]
8. Li, Y.J.; Lu, P.; Hu, D.Y.; Bhadury, P.S.; Zhang, Y.P.; Zhang, K.K. Determination of Dufulin Residue in Vegetables, Rice, and Tobacco Using Liquid Chromatography with Tandem Mass Spectrometry. *J. AOAC Int.* **2015**, *98*, 1739–1744. [CrossRef]
9. Bernardi, G.; Kemmerich, M.; Ribeiro, L.C.; Adaime, M.B.; Zanella, R.; Prestes, O.D. An effective method for pesticide residues determination in tobacco by GC-MS/MS and UHPLC-MS/MS employing acetonitrile extraction with low-temperature precipitation and d-SPE clean-up. *Talanta* **2016**, *161*, 40–47. [CrossRef]
10. Portoles, T.; Mol, J.G.J.; Sancho, J.V.; Lopez, F.J.; Hernandez, F. Validation of a qualitative screening method for pesticides in fruits and vegetables by gas chromatography quadrupole-time of flight mass spectrometry with atmospheric pressure chemical ionization. *Anal. Chim. Acta* **2014**, *838*, 76–85. [CrossRef]
11. Li, J.X.; Li, X.Y.; Chang, Q.Y.; Li, Y.; Jin, L.H.; Pang, G.F.; Fan, C.L. Screening of 439 Pesticide Residues in Fruits and Vegetables by Gas Chromatography-Quadrupole-Time-of-Flight Mass Spectrometry Based on TOF Accurate Mass Database and Q-TOF Spectrum Library. *J. AOAC Int.* **2018**, *101*, 1631–1638. [CrossRef]
12. Pang, G.F.; Fan, C.L.; Chang, Q.Y.; Li, J.X.; Kang, J.; Lu, M.L. Screening of 485 Pesticide Residues in Fruits and Vegetables by Liquid Chromatography-Quadrupole-Time-of-Flight Mass Spectrometry Based on TOF Accurate Mass Database and QTOF Spectrum Library. *J. AOAC Int.* **2018**, *101*, 1156–1182. [CrossRef]
13. Li, J.; Teng, X.; Wang, W.; Zhang, Z.; Fan, C. Determination of multiple pesticide residues in teas by gas chromatography with accurate time-of-flight mass spectrometry. *J. Sep. Sci.* **2019**, *42*, 1990–2002. [CrossRef]
14. Yang, F.; Bian, Z.; Chen, X.; Liu, S.S.; Liu, Y.; Tang, G. Determination of Chlorinated Phenoxy Acid Herbicides in Tobacco by Modified QuEChERS Extraction and High-Performance Liquid Chromatography/Tandem Mass Spectrometry. *J. AOAC Int.* **2013**, *96*, 1134–1137. [CrossRef]
15. Xiong, W.; Jing, H.; Guo, D.; Wang, Y.; Yang, F. A Novel Method for the Determination of Fungicide Residues in Tobacco by Ultra-performance Liquid Chromatography-Tandem Mass Spectrometry Combined with Pass-Through Solid-Phase Extraction. *Chromatographia* **2021**, *84*, 729–740. [CrossRef]
16. Li, M.; Jin, Y.; Li, H.-F.; Hashi, Y.; Ma, Y.; Lin, J.-M. Rapid determination of residual pesticides in tobacco by the quick, easy, cheap, effective, rugged, and safe sample pretreatment method coupled with LC-MS. *J. Sep. Sci.* **2013**, *36*, 2522–2529. [CrossRef]
17. Guo, W.; Bian, Z.; Zhang, D.; Tang, G.; Liu, W.; Wang, J.; Li, Z.; Yang, F. Simultaneous determination of herbicide residues in tobacco using ultraperformance convergence chromatography coupled with solid-phase extraction. *J. Sep. Sci.* **2015**, *38*, 858–863. [CrossRef]
18. Chen, Y.; Zhu, S.-C.; Zhen, X.-T.; Shi, M.-Z.; Yu, Y.-L.; Cao, J.; Zheng, H.; Ye, L.-H. Miniaturized solid phase extraction of multi-pesticide residues in food supplement using plant sorbent by microwave-induced activated carbons. *Microchem. J.* **2021**, *171*, 106814. [CrossRef]
19. Goon, A.; Shinde, R.; Ghosh, B.; Banerjee, K. Application of Automated Mini-Solid-Phase Extraction Cleanup for the Analysis of Pesticides in Complex Spice Matrixes by GC-MS/MS. *J. AOAC Int.* **2020**, *103*, 40–45. [CrossRef]
20. Amir, R.M.; Randhawa, M.A.; Nadeem, M.; Ahmed, A.; Ahmad, A.; Khan, M.R.; Khan, M.A.; Kausar, R. Assessing and Reporting Household Chemicals as a Novel Tool to Mitigate Pesticide Residues in Spinach (Spinacia oleracea). *Sci. Rep.* **2019**, *9*, 6. [CrossRef]
21. Cutillas, V.; Galera, M.M.; Rajski, L.; Fernandez-Alba, A.R. Evaluation of supercritical fluid chromatography coupled to tandem mass spectrometry for pesticide residues in food. *J. Chromatogr. A* **2018**, *1545*, 67–74. [CrossRef]
22. Mol, H.G.J.; Plaza-Bolanos, P.; Zomer, P.; de Rijk, T.C.; Stolker, A.A.M.; Mulder, P.P.J. Toward a Generic Extraction Method for Simultaneous Determination of Pesticides, Mycotoxins, Plant Toxins, and Veterinary Drugs in Feed and Food Matrixes. *Anal. Chem.* **2008**, *80*, 9450–9459. [CrossRef]
23. Lehotay, S.J. Determination of pesticide residues in foods by acetonitrile extraction and partitioning with magnesium sulfate: Collaborative study. *J. AOAC Int.* **2007**, *90*, 485–520. [CrossRef]
24. Chen, X.S.; Bian, Z.Y.; Tang, G.L.; Hu, Q.Y. Determination of 132 pesticide residues in tobacco by gas chromatography-tandem mass spectrometry. *Chin. J. Chromatogr.* **2012**, *30*, 1043–1055. [CrossRef]
25. Kwon, H.; Lehotay, S.J.; Geis-Asteggiante, L. Variability of matrix effects in liquid and gas chromatography-mass spectrometry analysis of pesticide residues after QuEChERS sample preparation of different food crops. *J. Chromatogr. A* **2012**, *1270*, 235–245. [CrossRef]
26. de Sousa, F.A.; Costa, A.I.G.; de Queiroz, M.; Teofilo, R.F.; Neves, A.A.; de Pinho, G.P. Evaluation of matrix effect on the GC response of eleven pesticides by PCA. *Food Chem.* **2012**, *135*, 179–185. [CrossRef]
27. Hakme, E.; Lozano, A.; Gomez-Ramos, M.M.; Hernando, M.D.; Fernandez-Alba, A.R. Non-target evaluation of contaminants in honey bees and pollen samples by gas chromatography time-of-flight mass spectrometry. *Chemosphere* **2017**, *184*, 1310–1319. [CrossRef]
28. Leandro, C.C.; Hancock, P.; Fussell, R.J.; Keely, B.J. Quantification and screening of pesticide residues in food by gas chromatography-exact mass time-of-flight mass spectrometry. *J. Chromatogr. A* **2007**, *1166*, 152–162. [CrossRef]
29. Lewis, R.J. *Sax's Dangerous Properties of Industrial Materials*, 11th ed; Wiley & Sons, Inc.: Hoboken, NJ, USA, 2004; p. 2592.
30. International Labour Office. *Encyclopedia of Occupational Health and Safety*; International Labour Office: Geneva, Switzerland, 1983; Volume I–II, p. 621.

Article

Simultaneous Determination of 23 Mycotoxins in Broiler Tissues by Solid Phase Extraction UHPLC-Q/Orbitrap High Resolution Mass Spectrometry

Youyou Yang [1,*,†], Zhuolin He [2,†], Lei Mu [3,4], Yunfeng Xie [3,4,*] and Liang Wang [2]

1. Institute of Animal Sciences of Chinese Academy of Agricultural Sciences, Beijing 100193, China
2. College of Life Science and Technology, Xinjiang University, Urumqi 830046, China; hzl863454690@163.com (Z.H.); WL1390593786@163.com (L.W.)
3. China Oil & Foodstuffs Corporation (COFCO) Nutrition and Health Research Institute, Beijing 102209, China; mulei1@cofco.com
4. Beijing Key Laboratory of Nutrition Health and Food Safety, Beijing 102209, China
* Correspondence: yangyou229@126.com (Y.Y.); xieyunfeng@cofco.com (Y.X.)
† These authors contributed equally to this work.

Citation: Yang, Y.; He, Z.; Mu, L.; Xie, Y.; Wang, L. Simultaneous Determination of 23 Mycotoxins in Broiler Tissues by Solid Phase Extraction UHPLC-Q/Orbitrap High Resolution Mass Spectrometry. *Separations* **2021**, *8*, 236. https://doi.org/10.3390/separations8120236

Academic Editor: Erica Liberto

Received: 22 October 2021
Accepted: 22 November 2021
Published: 4 December 2021

Publisher's Note: MDPI stays neutral with regard to jurisdictional claims in published maps and institutional affiliations.

Copyright: © 2021 by the authors. Licensee MDPI, Basel, Switzerland. This article is an open access article distributed under the terms and conditions of the Creative Commons Attribution (CC BY) license (https:// creativecommons.org/licenses/by/ 4.0/).

Abstract: Mycotoxins are a type of toxins harmful for not only animal but also human health. Cooccurrence of multi-mycotoxins could occur for food infected by several molds, producing multi-mycotoxins. It is necessary to develop corresponding determination methods, among which current mass spectrometry (MS) dominates. Currently, the accurate identification and quantitation of mycotoxins in complex matrices by MS with low resolution is still a challenge since false-positive results are typically obtained. Here, a method for the simultaneous determination of 23 mycotoxins in broiler tissues using ultra-high performance liquid chromatography-quadrupole/orbitrap HRMS was established. After the extraction by acetonitrile-water-formic acid (80:18:2, $v/v/v$), the purification by multifunctional purification solid phase extraction cartridges and the chromatographic separation on a C18 column, representative mycotoxins were determined by HRMS in full scan/data-dependent MS/MS acquisition mode. The quantitation was based on the external standard method. An MS/MS database of 23 mycotoxins was established to achieve qualitative screening and simultaneous quantification. Mycotoxins had a good linear relationship within a certain concentration range with correlation coefficients (r^2) larger than 0.991 as well as the limit of quantitation of 1.80–300 µg/kg. The average recoveries at three different levels of low, medium and high fortification were 61–111% with relative standard deviations less than 13.5%. The method was fast, accurate, and suitable for the precise qualification of multiple mycotoxins in broiler tissues. 15 µg/kg zearalenone (ZEN) was detected in one liver sample among 30 samples from markets including chicken breast meat, liver, and gizzards. The result illustrated that the pollution of ZEN should not be neglected considering its harmful effect on the target organ of liver.

Keywords: broiler tissue; orbitrap high resolution mass spectrometry; mycotoxins; rapid screening; solid phase extraction

1. Introduction

Mycotoxins are secondary metabolites with low molecular weight, approximately of <1000 Da, produced by fungus species during growth and proliferation. The corresponding classification is difficult due to the complex structures and origins. Mycotoxins include group 1 and group 2B carcinogens (for example, aflatoxins (AFs)) are considered as the most toxic. In addition, there also exist modified and emerging mycotoxins. Mycotoxins which are harmful for animal liver and decrease animal' immunity and reproduction capacity can enter animal derived foods including meat, egg and milk and lead to residue through food chain, storage and processing, severely threatening human and animal health [1,2]. Currently, there are more than 400 mycotoxins, only a few of which are daily regulated and routinely monitored [3]. The current research about the residue determination of mycotoxins paid more attention to feed, grain, and oil to prevent them from entering the food chain. However, the detection of mycotoxins in meat (especially animal derived organs) was seldom reported. It was determined that the elimination time of mycotoxin residue in liver and muscle reached the quantitative value within at least 11–18 d [4]. Zearalenone (ZEN) in chicken serum was completely eliminated after 7 d with oral administration of the feed polluted by mycotoxins. However, there existed ZEN in both liver and faeces [5,6]. Mycotoxin residue in liver, kidney, muscle and milk of animal derived from food was mainly due to oral intake of the feed polluted by mycotoxins. Through food chain, mycotoxins entered the human body, threatening the human health [7–9]. Mycotoxin pollution which was widespread and uncontrollable has become a crucial aspect of animal derived food safety.

Over the last two decades, considering the determination of mycotoxins, MS including tandem MS and HRMS dominated (as high as 55%). The determination methods of mycotoxins mainly included HPLC [10,11], GC [12,13], and LC-MS/MS [14–19]. Tandem MS for mycotoxins has been widely studied, focusing on multi-mycotoxin residue analysis and quantitation which was "golden standard". Tandem MS such as triple quadrupole (QQQ) of unit resolution MS selected multiple reaction monitoring (MRM) for the quantitation. The ion transitions and related parameters were necessary to be optimized sequentially, which were labor and time consuming when aiming at a large number of targets. The unit resolution MS was susceptible to the isobaric ion, allowing potential false positive phenomenon [19].

Recently, HRMS including Q/orbitrap and time of flight has been utilized in the determination of mycotoxins, which mainly focused on method development for avoiding matrix interference and accurate identification. Previous research work was more focused on the determination of single or a type of mycotoxins with a relative narrow covering range. Q/orbitrap HRMS has lots of advantages such as high resolution and high accuracy, which can realize accurate screening. It can obtain accurate molecular weights of compounds as well as the fragment ions under high resolution and with relatively strong anti-interference capacity [20,21]. Its most attractive advantage is the feasibility with target, non-target, and retrospective analysis. The current reported research about the determination of mycotoxins in meat has been mainly based on the unit resolution MS. The combination of UHPLC-Q/orbitrap HRMS has been utilized in the determination of pollutants in animal derived food [21] and pesticide residue [22].

In addition, the pretreatment method of mycotoxins has been mainly coupled with tandem MS. The corresponding coupling with HRMS was seldom. Currently, only QuEChERS method has been coupled to HRMS. However, fumonisins (FBs) which were known to be difficult to analyze with QuEChERS, were not considered by HRMS [23].

Considering the dietary habit of eating chicken meat as well as liver and gizzard in China, it was crucial to develop methods for analyzing mycotoxins in different broiler tissues and organs. In this work, A total of 23 mycotoxins including AFs (AFTB$_1$, AFTB$_2$, AFTG$_1$, AFTG$_2$, AFTM$_1$, AFTM$_2$), deoxynivalenol (DON), 3-acetyldeoxynivalenol (3-ADON), 15-acetyldeoxynivalenol (15-ADON), de-epoxydeoxynivalenol (DOM), T-2 toxin (T-2), HT-2 toxin (HT-2), FBs (FB$_1$, FB$_2$, FB$_3$), ochratoxin A (OTA), ochratoxin (OTB), zearalanone (ZAN), zearalenone (ZEN), α-zearalanol (α-ZAL), β-zearalanol (β-ZAL), α-zearalenol (α-ZOL) and β-zearaalenol (β-ZOL) in broiler tissues were determined by solid phase extraction (SPE)-UHPLC-Q/orbitrap HRMS for rapid and accurate identification and quantitation, providing supports for animal derived food safety.

2. Materials and Methods

2.1. Reagents

AFTB$_1$ (2 µg/mL), AFTB$_2$ (0.5 µg/mL), AFTG$_1$ (2 µg/mL), AFTG$_2$ (0.5 µg/mL), AFTM$_1$ (0.5 µg/mL), AFTM$_2$ (0.5 µg/mL), FB$_1$ (50 µg/mL), FB$_2$ (50 µg/mL), FB$_3$ (50 µg/mL),

DON (100 µg/mL), DOM (25 µg/mL), 3-ADON (100 µg/mL), 15-ADON (100 µg/mL), T-2 toxin (100 µg/mL), HT-2 toxin (100 µg/mL), STC (50 µg/mL), OTA (10 µg/mL), OTB (10 µg/mL) were purchased from Romer Labs Co., Ltd. (Tulln, Austria). ZAN (100 µg/mL), ZEN (100 µg/mL), α-ZAL (100 µg/mL), α-ZEL (100 µg/mL), β-ZOL (100 µg/mL), and β-ZOL (100 µg/mL) were purchased from Anpel Co., Ltd. (Shanghai, China). Chicken breast, liver and kidneys were purchased from local super market.

Methanol and acetonitrile were all HPLC grade and purchased from Fisher Scientific (Pittsburgh, PA, USA). Water was purified by a Milli Q Advantaged A10 water purification system (Millipore, Bedford, MA, USA). Formic acid for UPLC/LC-MS were from Anpel Co., Ltd. (Shanghai, China). SPE columns with Captiva-EMR Lipid (600 mg, 6 mL,) and Oasis PRIME HLB (600 mg, 6 mL) were from Agilent Technologies (Santa Clara, CA, USA) and Waters (Shanghai, China).

2.2. Instrument Conditions

Separation and detection of mycotoxins performed in the Q-Exactive system combined with Ultimate 3000 LC (Thermofisher, San Jose, CA, USA). Separation was fulfilled using a CORTECS C18 column (2.1 × 100 mm, 1.6 µm; Waters, Wexford, Ireland). The injection volume was set at 10 µL and the flow rate was maintained at 0.2 mL/min. The mobile phase was composed of water (0.1% formic acid) as eluent A and methanol as eluent B. The gradient elution program was shown in Table 1. The mass spectrometer was equipped with a heated electrospray ionization (H-ESI) source. Data with positive and negative modes were acquired through data-dependent acquisition, respectively. The mass spectrometer parameters were as follows: spray voltage, 3000 V (±); auxiliary gas heater temperature, 350 °C; capillary temperature, 320 °C; sheath gas flow rate 40 Arb; auxiliary gas flow rate, 15 Arb; scan range, 50–600 m/z; collision energy (NCE): 20, 30, 40 V; the resolving power for MS1 and MS2, 70,000 and 17,500, respectively.

Table 1. Gradient elution programs.

Acquisition Mode	Time (min)	Gradient (%) A	Gradient (%) B	Acquisition Mode	Time (min)	Gradient (%) A	Gradient (%) B
Positive mode	1	70	30	Negative mode	1	97	3
	6.5	45	55		2	45	55
	8.5	45	55		9	30	70
	10	20	80		10	1	99
	12	20	80		11	1	99
	12.1	70	30		11.1	97	3
	16.1	70	30		15	97	3

2.3. Sample Preparation

Weigh 2.00 ± 0.05 g of the minced sample into a 50 mL centrifuge tube. 10 mL acetonitrile/water/formic acid (80/18/2, $v/v/v$) were added. The mixture was vortexed for 1 min and processed the ultrasonication at 30 °C for 20 min, followed by centrifugation at 8000 rpm for 10 min. 5 mL of the supernatant was transferred to the Captiva-EMR Lipid and controlled at the eluting rate of 3 drops/s. Right after the solution flowed through the column, 1 mL acetonitrile/water (80/20, v/v) was added. Both of the elution solutions were collected and nearly dried through the nitrogen flow at 40 °C. The dried eluents were redissolved in a mixture of methanol/water/formic acid (250 µL, 30/70/0.1, $v/v/v$), vortexed for 1 min, ultrasonicated for 5 min, and centrifuged at 12,000 rpm for 10 min. The supernatant was transferred into a vial for analysis.

2.4. Method Validation

Linearity, sensitivity, accuracy, and precision were investigated according to Criterion on quality control of laboratories—chemical testing of food (GB 27404). Calibration curves were constructed through the responses versus the concentrations spiked in the blank matrix. The limit of detection (LOD, S/N = 3) and the limit of quantification (LOQ, S/N = 10) were calculated in light of the blank matrix with the lowest spiking level. Recoveries and stability were investigated using blank samples fortified with three different levels. Samples of each level were prepared in six replicates. The recovery of each mycotoxin was calculated as the ratio of the mean peak areas between the samples spiked before extraction and the samples spiked after extraction. The relative standard deviation (RSD) of peak areas in six replicates for each mycotoxin at three spiking levels represented the stability of the method.

3. Results and Discussion

3.1. The Optimization of LC-HRMS Conditions

In the optimization procedure of the chromatography conditions, the column type and the mobile phase were investigated. Compared with Thermo Scientific Accucore C18 column (2.1 mm × 100 mm, 2.6 µm), the separation efficiency of CORTECS-UPLC-C18 (2.1 mm × 100 mm, 1.6 µm) was higher, which could guarantee the good peak shape of each target compound. Besides, the analysis time was shorter and the resolution was relatively higher. Thus, analysis of 23 mycotoxins could be fulfilled within a shorter time. In this experiment, methanol was selected as the organic phase. Modifiers of 0.1% formic acid and 5 mmol/L ammonium formate in aqueous phase were investigated. The addition of formic acid in aqueous phase resulted in better responses of target mycotoxins since the formic acid could facilitate the protonization of some mycotoxins. Thus, methanol-0.1% aqueous formic acid (v/v) was chosen as the mobile phase. The extracted ion chromatograms of mycotoxins were shown in Figure 1.

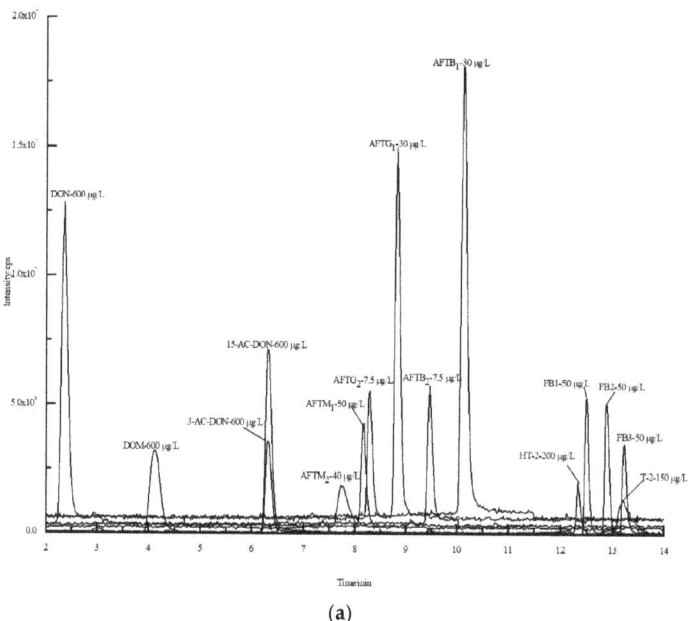

(a)

Figure 1. Cont.

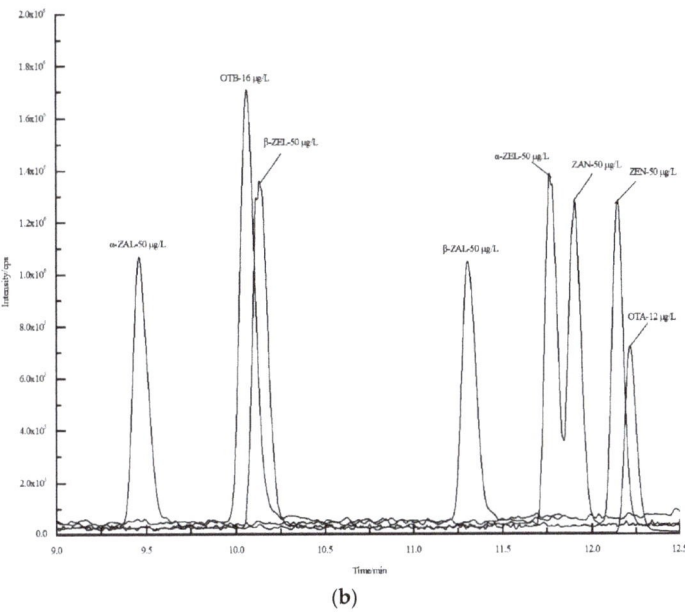

(b)

Figure 1. Extracted ion chromatograms of the 23 mycotoxins at positive mode (**a**) and negative mode (**b**).

Different spray voltages of 2.8, 3.0, and 3.5 kV for heated ESI (HESI) were studied for the ionization of targets, illustrating that 3.0 kV led to better ionization efficiency of target compounds. According to accurate identification of HRMS, two ions of mass tolerance within 5 ppm are required. One of these ions is required to be a fragment ion while the second ion should be the (de-) protonated molecular ion or an adduct ion thereof. In this work, full MS/data dependent-MS^2 (full MS/dd-MS^2) mode would be chosen and would give the information of precursor ions and MS/MS spectra. Retention time, accurate m/z of molecular ion and accurate m/z of the fragment were listed in Table 2. Within 14.0 min, satisfactory separation and detection were realized. According to the structure and the properties of mycotoxins, both positive and negative ionization modes were used.

3.2. The Optimization of SPE Pretreatment

The extraction solvent directly influenced recoveries of compounds. To obtain higher recoveries and decrease matrix interference, considering characteristics of protein and lipid contents in broiler tissue, different extraction solutions were optimized in order to realize simultaneous extraction of 23 mycotoxins. Methanol/formic acid (98:2, v/v), acetonitrile/formic acid (98:2, v/v) and acetonitrile/water/methanol (80:18:2, $v/v/v$) were investigated for preparation of broiler tissues including chicken breast meat, gizzards, and liver. When acidified methanol was used, two phases of solid-liquid could not be well separated, and the extraction solution was also in the muddy state even after centrifugation. However, no such phenomenon would happen when acetonitrile was used. Thus, acidified acetonitrile was used to extract 23 mycotoxins. The recovery comparison of two extraction solutions were shown in Figure 2. Finally, acetonitrile/water/methanol (80:18:2, $v/v/v$) was selected.

Table 2. Qualification parameters of 23 mycotoxins.

Comment	Ion Mode	Measured Mass (m/z)	Characteristic Ion 1 (m/z)	Characteristic Ion 2 (m/z)	RT/min
AFB$_1$	[M + H]$^+$	313.07111	285.07611	270.05267	9.68
AFB$_2$	[M + H]$^+$	315.08661	287.09186	259.06024	9.22
AFG$_1$	[M + H]$^+$	329.06577	243.06554	283.06055	8.42
AFG$_2$	[M + H]$^+$	331.08191	313.07135	245.08141	7.86
AFM$_1$	[M + H]$^+$	329.06577	273.07593	259.06036	7.68
AFM$_2$	[M + H]$^+$	331.08008	273.07608	285.07596	6.70
T-2	[M + NH$_4$]$^+$	484.25464	305.13736	185.09566	12.34
HT-2	[M + NH$_4$]$^+$	442.24233	263.12665	235.10591	11.90
FB$_1$	[M + H]$^+$	722.39337	704.38312	352.32013	11.75
FB$_2$	[M + H]$^+$	706.3985	336.32513	688.38812	13.15
FB$_3$	[M + H]$^+$	706.3985	336.32523	688.38812	12.82
DON	[M + H]$^+$	297.13287	249.11194	203.10658	2.35
DOM	[M + H]$^+$	281.13724	235.10661	137.05975	4.18
15-ADON	[M + H]$^+$	339.14368	323.12293	137.05972	6.28
3-ADON	[M + H]$^+$	339.14368	231.10149	279.12253	6.28
α-ZAL	[M − H]$^-$	323.17032	277.18048	303.15970	9.45
β-ZAL	[M − H]$^-$	323.17041	277.1806	303.15982	11.29
α-ZOL	[M − H]$^-$	319.15454	275.16489	160.01656	11.76
β-ZOL	[M − H]$^-$	319.15463	275.16495	160.01651	10.13
ZAN	[M − H]$^-$	319.1546	275.16501	205.08682	11.91
ZEN	[M − H]$^-$	317.13907	131.05017	175.03992	12.14
OTA	[M − H]$^-$	402.07407	358.08435	231.01634	12.23
OTB	[M − H]$^-$	368.11105	324.12436	280.09824	10.06

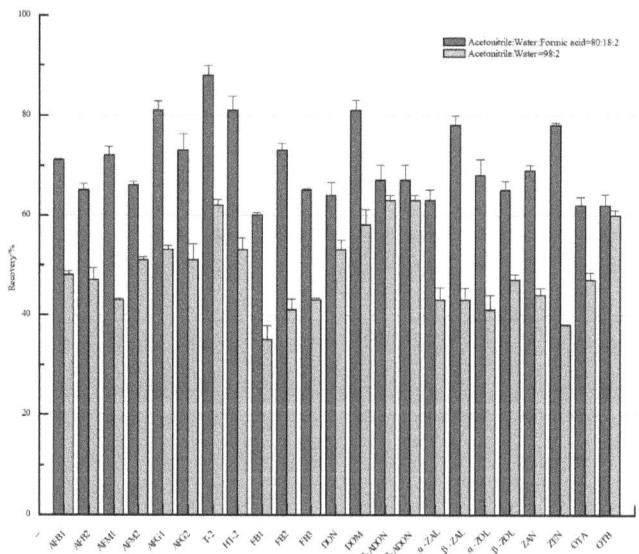

Figure 2. The effect of different extracts on the recovery of each mycotoxin.

Chicken breast meat which was cheap and easily available were widely consumed. Chicken liver and gizzard were the characteristic Chinese food. Considering 1.4 billion Chinese people, the method development was of great importance. As for different matrices of chicken breast meat, liver and gizzard, a great certain of proteins, lipids and minerals would be retained in the extraction solution, which brought in interference to mycotoxin determination. Besides, different mycotoxins would have different physicochemical properties. Therefore, it would be of great necessity to develop the method which would be suitable for multiple mycotoxins. In this study, two multifunctional purification cartridges were investigated for the purification effect of 23 mycotoxins, such as Oasis PRIME-HLB and Captiva-EMR Lipid purification cartridges, whose purification effects on mycotoxin recoveries were shown in Figure 3. Both of those two columns allowed direct sample loading without equilibrium and activation and belonging to filtering SPE cartridge for the impurity removal, which greatly simplified the procedures of SPE and effectively decreased sample treatment time [24,25]. Captiva-EMR Lipid purification cartridge also had a higher absorptive capacity and removal efficiency of lipids. According to the response after the purification by Captiva EMR Lipid and Oasis PRIME-HLB cartridges, recoveries of 23 mycotoxins with the pretreatment of Captiva EMP Lipid cartridge were acceptable, namely 61–111%, wherein FBs also demonstrated good results. Thus, in this work, the Captiva-EMR Lipid purification cartridge was chosen.

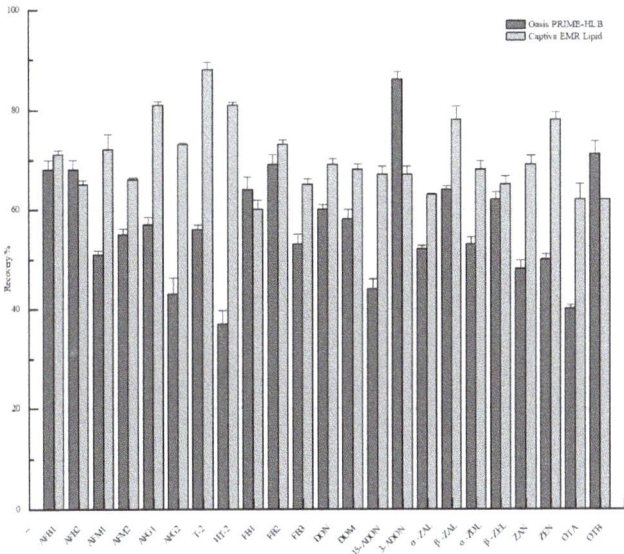

Figure 3. The influence of different purification columns on the recovery of each mycotoxin.

3.3. Matrix Effect

The matrix effect was mainly due to endogenous components in the sample as well as impurities introduced in the pretreatment process. In ESI, the response of different mycotoxins was easily influenced by matrices, which was represented by the slope ratio between linear regression equation in blank matrix and that in the solvent. As shown in Table 3, most of the mycotoxins demonstrated relatively strong matrix effect. Thus, the matrix-matched linearity was used in this study to make up for the influence of matrix effect, guaranteeing the stability and accuracy of MS results.

3.4. The Method Validation

The recovery and repeatability assays were processed in three different blank matrices (chicken breast meat, liver and gizzard), respectively. And they were investigated at three spiking levels in six replicates. As shown in Table 4, recoveries of 23 mycotoxins were between 61% and 111% with RSDs < 15%. The good recovery and repeatability verified that the established SPE-UHPLC Q/orbitrap HRMS method was suitable for routine risk monitoring of 23 mycotoxins in broiler tissues such as breast meat, liver, and gizzards.

Table 3. Evaluation of matrix effect.

Mycotoxins	Matrix Effect/%		
	Breast	Gizzard	Liver
AFB$_1$	88.64	71.29	59.10
AFB$_2$	88.14	40.46	55.95
AFG$_1$	90.35	62.03	85.96
AFG$_2$	78.85	66.55	70.11
AFM$_1$	80.98	59.78	87.02
AFM$_2$	87.94	55.50	87.72
T-2	48.11	32.62	42.23
HT-2	46.24	35.27	46.98
FB$_1$	15.96	16.00	46.82
FB$_2$	38.38	20.79	54.17
FB$_3$	48.29	73.76	39.81
DON	34.31	10.32	36.97
DOM	90.11	76.82	67.87
15-ADON	87.14	63.23	57.97
3-ADON	87.14	58.27	62.28
α-ZAL	21.54	40.17	23.75
β-ZAL	36.61	46.91	48.26
α-ZOL	32.47	47.98	37.62
β-ZOL	37.78	57.52	12.29
ZAN	40.42	26.37	50.35
ZEN	44.22	44.32	44.01
OTA	30.35	32.43	58.76
OTB	49.42	32.21	29.26

Table 4. Recoveries of 23 mycotoxins at 3 levels (n = 6).

Mycotoxins	Added Concentration (μg/kg)	Chicken Liver		Chicken Gizzard		Chicken Breast Meat	
		Recovery/%	RSD/%	Recovery/%	RSD/%	Recovery/%	RSD/%
AFB$_1$	7.5	85	0.3	64	0.8	67	10.1
	15	67	5.4	65	1.7	79	9.0
	37.5	69	11.9	69	0.4	80	5.0
AFB$_2$	1.875	68	1.3	62	2.4	101	3.6
	3.75	65	2.6	68	3.2	77	10.9
	9.375	73	7.6	81	1.4	69	3.9
AFG$_1$	7.5	61	1.8	84	0.3	73	2.8
	15	62	3.4	71	1.5	84	8.6
	37.5	61	1.7	61	1.7	70	8.6
AFG$_2$	1.875	69	0.8	67	0.6	67	1.3
	3.75	67	4.1	61	1.3	72	10.5
	9.375	63	4.1	66	12.5	73	5.5
AFM$_1$	12.5	67	1.7	70	0.9	64	2.0
	25	92	3.1	66	5.1	65	8.1
	62.5	65	3.3	68	3.2	63	4.1
AFM$_2$	10	78	3.3	87	6.2	63	2.0
	20	91	1.3	89	8.9	62	2.0
	50	88	5.4	82	3.6	66	6.1
T-2	37.5	75	5.9	80	9.2	70	10.9
	75	97	9.4	67	12.6	67	9.0
	187.5	67	2.4	67	4.1	75	4.0

Table 4. Cont.

Mycotoxins	Added Concentration (µg/kg)	Chicken Liver		Chicken Gizzard		Chicken Breast Meat	
		Recovery/%	RSD/%	Recovery/%	RSD/%	Recovery/%	RSD/%
HT-2	52.5	87	2.7	75	2.4	87	2.3
	105	75	8.2	88	4.1	71	13.5
	262.5	93	3.1	86	3.4	72	3.5
FB$_1$	12.5	69	0.5	62	2.9	66	3.4
	25	73	3.0	77	1.2	65	7.2
	62.5	75	6.9	81	6.2	77	4.2
FB$_2$	12.5	102	1.4	66	2.2	64	8.0
	25	68	3.9	67	3.6	64	6.5
	62.5	87	2.2	86	6.0	72	7.5
FB$_3$	12.5	64	4.4	65	1.4	64	6.9
	25	76	9.5	69	0.3	65	9.7
	62.5	66	5.1	64	4.9	65	7.7
DON	350	67	1.2	64	2.6	66	1.1
	700	68	2.8	67	3.8	68	7.9
	1750	67	2.7	69	4.6	76	5.8
DOM	300	65	1.8	66	1.3	65	1.4
	600	67	3.6	81	1.9	81	7.8
	1500	75	1.1	75	2.1	69	1.0
15-ADON	300	63	3.1	69	1.1	94	9.2
	600	68	7.9	74	3.1	111	5.3
	1500	68	3.2	69	5.3	71	4.3
3-ADON	300	63	3.1	69	1.1	93	3.1
	600	68	7.9	73	3.1	111	5.8
	1500	68	3.2	69	5.3	78	3.8
α-ZAL	12.5	95	2.1	77	2.5	67	7.3
	25	64	5.3	87	1.1	86	5.1
	62.5	86	1.6	88	2.7	63	4.1
β-ZAL	12.5	94	2.0	76	2.4	65	6.6
	25	66	5.3	87	1.1	80	3.8
	62.5	83	1.5	88	2.5	62	3.8
α-ZOL	12.5	86	3.2	83	3.0	62	4.5
	25	64	2.3	85	1.3	72	5.5
	62.5	66	2.0	81	1.8	64	6.5
β-ZOL	12.5	71	1.9	79	3.1	66	8.4
	25	75	9.4	95	1.0	83	6.7
	62.5	87	3.4	95	0.8	65	5.7
ZAN	12.5	76	5.1	85	2.4	64	7.7
	25	71	10.0	69	2.0	76	5.5
	62.5	72	4.0	69	1.9	62	3.5
ZEN	12.5	84	0.6	78	0.2	64	5.0
	25	96	2.1	68	0.4	77	3.6
	62.5	79	8.2	71	0.6	68	4.6
OTA	3	83	3.7	74	3.5	70	6.5
	6	69	3.7	102	2.3	85	5.5
	15	95	5.1	61	1.4	66	7.5
OTB	3.75	66	2.3	66	1.1	77	6.1
	7.5	66	8.1	73	0.9	72	5.7
	18.75	70	5.9	71	0.7	89	4.7

Mixed standard solutions of mycotoxins with different concentrations were obtained in blank matrix solution. The linear regression equation was plotted through y of peak area and x of concentration (μg/L) as shown in Table 5. Linearity of 23 mycotoxins with correlation coefficients (R^2) larger than 0.991 was obtained. LODs were in the range of 0.40–130.00 μg/kg and LOQs ranged from 1.20 to 350.00 μg/kg in those three matrices.

3.5. The Real Sample Analysis

With the optimized method, 30 samples including breast meat, liver and gizzard purchased from different markets were determined. ZEN was detected in one of chicken liver samples with the concentration of 15.29 μg/kg. Other samples' determination results were below the corresponding LOD. The MS/MS spectra of ZEN in the contaminated liver sample was shown in Figure 4. The occurrence of the positive result was due to the polluted feeds, which resulted in mycotoxin residue in animal. In addition, the pollution during the processing, storage, and marketing processes of meat was also possible.

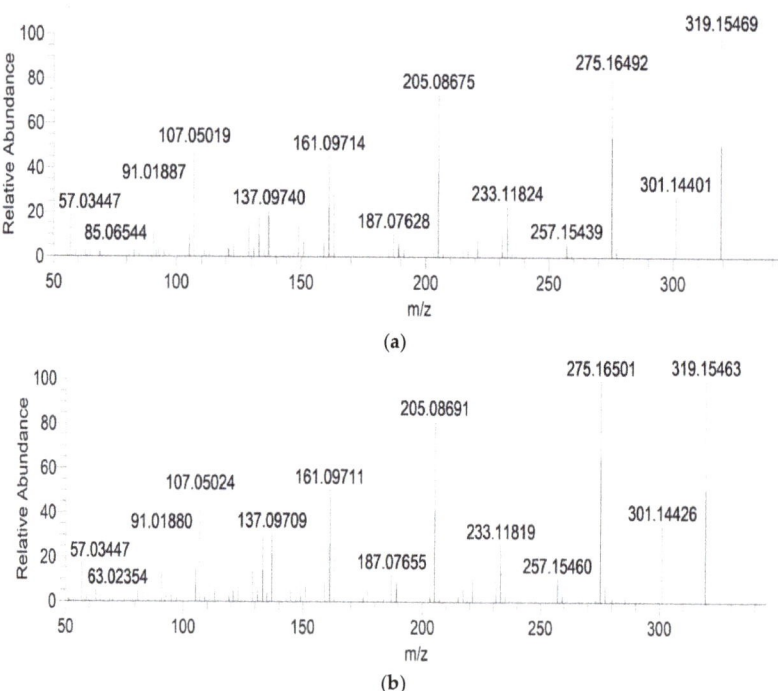

Figure 4. MS^2 spectra of ZEN in the standard solution (**a**) and contaminated sample (**b**).

ZEN was one of the most important mycotoxins, which altered fertility and reproduction and influenced hepatic cellular immune response. After oral administration, ZEN was difficult to be detected in vivo, which was not appropriate as the biomarker. However, in this work, ZEN's concentration in the positive liver sample was relatively high, illustrating that liver was the target organ reported in some publications. Besides, in the risk assessment report of ZEN from European Food Scientific Agency (EFSA) in 2011, the tolerable daily intake (TDI) was 0.25 μg/kg body weight. According to the adult body weight of 60 kg, although the ZEN intake of this positive liver sample didn't exceed the standard, the long-term or large dose intake hazards had to be worried especially considering those susceptible groups such as pregnant women or children.

Table 5. Linear range, detection limit and quantification limit of 23 mycotoxins.

Mycotoxins	Chicken Breast					Chicken Gizzard					Chicken Liver				
	Linear Range (µg/L)	Linear Equation	Correlation Coefficient (R2)	LOD (µg/kg)	LOQ (µg/kg)	Linear Range (µg/L)	Linear Equation	Correlation Coefficient (R2)	LOD (µg/kg)	LOQ (µg/kg)	Linear Range (µg/L)	Linear Equation	Correlation Coefficient (R2)	LOD (µg/kg)	LOQ (µg/kg)
AFB$_1$	15–600	$Y = 1.774 \times 10^7 X + 1.826 \times 10^8$	0.994	2.5	7.5	10–400	$Y = 6.18 \times 10^7 X - 5.973 \times 10^7$	0.994	2	6	10–400	$Y = 1.115 \times 10^7 X + 1.125 \times 10^8$	0.99	1.6	5
AFB$_2$	3.75–150	$Y = 3.381 \times 10^6 X + 2.306 \times 10^6$	0.999	0.6	1.8	2.5–100	$Y = 5.679 \times 10^6 X - 1.568 \times 10^7$	0.995	0.5	1.5	2.5–100	$Y = 2.18 \times 10^6 X - 5.715 \times 10^6$	0.999	0.4	1.2
AFG$_1$	15–600	$Y = 1.189 \times 10^7 X + 4.353 \times 10^6$	0.999	2.5	7.5	10–400	$Y = 7.338 \times 10^6 X + 1.415 \times 10^7$	0.998	2	6	10–400	$Y = 8.478 \times 10^6 X + 7.596 \times 10^7$	0.993	2	6
AFG$_2$	3.75–150	$Y = 2.401 \times 10^7 X - 3.781 \times 10^4$	0.999	0.6	1.8	2.5–100	$Y = 7.178 \times 10^7 X + 1.472 \times 10^7$	0.995	0.5	1.5	2.5–100	$Y = 8.034 \times 10^6 X + 1.568 \times 10^7$	0.994	0.5	1.5
AFM$_1$	25–1000	$Y = 1.189 \times 10^7 X + 2.323 \times 10^7$	0.999	4.1	12.5	12.5–500	$Y = 2.956 \times 10^7 X - 1.481 \times 10^7$	0.997	3	9	25–1000	$Y = 9.161 \times 10^6 X - 3.877 \times 10^7$	0.998	4.1	12.5
AFM$_2$	20–800	$Y = 3.995 \times 10^6 X + 3.346 \times 10^6$	0.999	3.33	10	10–400	$Y = 8.977 \times 10^6 X - 9.662 \times 10^4$	0.998	2.6	8	20–800	$Y = 3.253 \times 10^6 X - 1.48 \times 10^7$	0.999	3.33	10
T-2	75–3000	$Y = 5.261 \times 10^6 X - 8.745 \times 10^5$	0.999	12.5	37.5	37–1500	$Y = 3.966 \times 10^6 X + 6.703 \times 10^6$	0.999	11	33	75–3000	$Y = 3.699 \times 10^6 X - 3.321 \times 10^6$	0.99	12.5	37.5
HT-2	105–4200	$Y = 6.596 \times 10^5 X - 4.328 \times 10^5$	0.999	17.5	52.5	52–2100	$Y = 1.019 \times 10^6 X - 9.894 \times 10^5$	0.992	16	48	105–4200	$Y = 1.244 \times 10^6 X - 1.584 \times 10^6$	0.999	17.5	52.5
HB$_1$	12.5–500	$Y = 1.496 \times 10^7 X - 2.147 \times 10^7$	0.997	3	9	12.5–500	$Y = 1.781 \times 10^7 X - 2.425 \times 10^7$	0.99	3	9	25–1000	$Y = 2.81 \times 10^7 X - 1.435 \times 10^7$	0.995	4.17	12.5
HB$_2$	12.5–500	$Y = 1.239 \times 10^7 X - 1.408 \times 10^7$	0.997	3	9	12.5–500	$Y = 1.539 \times 10^7 X - 1.507 \times 10^7$	0.994	3.3	10	25–1000	$Y = 2.66 \times 10^6 X - 1.205 \times 10^7$	0.992	4.17	12.5
HB$_3$	12.5–500	$Y = 1.781 \times 10^7 X - 1.608 \times 10^7$	0.999	3	9	12.5–500	$Y = 4.025 \times 10^7 X + 2.219 \times 10^7$	0.994	3.3	10	25–1000	$Y = 2.959 \times 10^6 X + 941 \times 10^6$	0.997	4.17	12.5
DON	700–28,000	$Y = 1.773 \times 10^5 X + 1.106 \times 10^8$	0.991	130	350	350–14,000	$Y = 1.986 \times 10^5 X + 1.826 \times 10^8$	0.993	100	300	700–28,000	$Y = 2.81 \times 10^5 X + 3.715 \times 10^7$	0.994	130	350
D3M	600–24,000	$Y = 7.032 \times 10^5 X + 4.598 \times 10^8$	0.996	100	300	300–12,000	$Y = 3.986 \times 10^5 X + 2.436 \times 10^8$	0.995	93	280	300–12,000	$Y = 4.19 \times 10^5 X + 2.715 \times 10^8$	0.991	93	280
15-AC-DON	600–24,000	$Y = 3.089 \times 10^6 X + 1.774 \times 10^8$	0.999	100	300	300–12,000	$Y = 8.444 \times 10^6 X + 3.126 \times 10^7$	0.993	93	280	300–12,000	$Y = 7.35 \times 10^6 X + 5.715 \times 10^7$	0.992	93	280
3-AC-DON	600–24,000	$Y = 3.089 \times 10^6 X + 1.774 \times 10^8$	0.999	100	300	300–12,000	$Y = 7.544 \times 10^6 X + 3.126 \times 10^7$	0.999	93	280	300–12,000	$Y = 8.19 \times 10^6 X + 5.715 \times 10^7$	0.996	93	280
α-ZAL	12.5–500	$Y = 1.464 \times 10^7 X - 4.241 \times 10^7$	0.993	3.75	11	25–1000	$Y = 2.452 \times 10^7 X - 2.507 \times 10^7$	0.99	4.1	12.5	12.5–500	$Y = 1.183 \times 10^7 X + 1.52 \times 10^6$	0.999	3.75	11
β-ZAL	12.5–500	$Y = 1.832 \times 10^7 X - 5.799 \times 10^7$	0.991	3	9	25–1000	$Y = 3.451 \times 10^7 X - 2.62 \times 10^7$	0.99	4.1	12.5	12.5–500	$Y = 3.541 \times 10^7 X - 2.079 \times 10^7$	0.996	3.75	11
α-ZEL	12.5–500	$Y = 2.086 \times 10^7 X - 6.629 \times 10^7$	0.993	3	9	25–1000	$Y = 4.01 \times 10^7 X - 1.722 \times 10^7$	0.999	4.1	12.5	12.5–500	$Y = 3.344 \times 10^6 X - 2.162 \times 10^7$	0.993	3.75	11
β-ZEL	12.5–500	$Y = 1.891 \times 10^7 X - 5.887 \times 10^7$	0.992	3	9	25–1000	$Y = 4.45 \times 10^7 X - 1.826 \times 10^{26}$	0.994	4.1	12.5	12.5–500	$Y = 2.156 \times 10^7 X - 3.531 \times 10^7$	0.991	3.75	11
ZAN	12.5–500	$Y = 2.555 \times 10^7 X - 7.456 \times 10^7$	0.993	3	9	25–1000	$Y = 3.474 \times 10^7 X - 1.004 \times 10^8$	0.995	4.1	12.5	12.5–500	$Y = 5.146 \times 10^7 X - 2.079 \times 10^6$	0.997	3.75	11
ZEN	12.5–500	$Y = 2.361 \times 10^7 X - 6.68 \times 10^7$	0.994	3	9	25–1000	$Y = 1.636 \times 10^7 X - 5.971 \times 10^7$	0.9958	4.1	12.5	12.5–500	$Y = 4.224 \times 10^7 X - 3.686 \times 10^6$	0.9956	3.75	11
OTA	3–240	$Y = 4.359 \times 10^6 X - 7.068 \times 10^6$	0.992	0.6	2	3–240	$Y = 6.459 \times 10^6 X - 4.288 \times 10^6$	0.9989	0.6	2	6–480	$Y = 1.057 \times 10^7 X - 8.678 \times 10^{25}$	0.9926	1	3
OTB	7.5–600	$Y = 4.373 \times 10^6 X - 1.36 \times 10^7$	0.991	1.25	3.75	3.75–300	$Y = 6.785 \times 10^6 X - 1.826 \times 10^6$	0.9943	1	3	3.75–300	$Y = 3.38 \times 10^6 X - 5.715 \times 10^{26}$	0.9918	1	3

Thus, to ensure the quality of broiler tissue, besides the safeguard of broiler feeds, the possible mycotoxin pollution of broiler in the slaughter or storage process should also be paid attention to.

4. Conclusions

An SPE-UHPLC Q/orbitrap method was established for rapid screening, accurate identification and quantitation of 23 mycotoxins in broiler tissues. The screening database of 23 mycotoxins was established, including retention time, accurate precursor m/z and MS/MS fragment, and could facilitate the rapid screening of mycotoxins. In this work, new multi-functional EMR column was applied in the removal of the lipids. Good recoveries were obtained and were in the range of 61–111%. In addition, high accuracy and high anti-interference capacity could be achieved in this method. ZEN at the concentration of 15 μg/kg was detected in one liver sample among 30 real broiler meat and organ samples, showing the possible harmful effect on the target organ of liver.

Author Contributions: Conceptualization, Y.Y. and Y.X.; Methodology, Y.Y., Z.H. and L.M.; Project administration, Y.X. and Y.Y.; Supervision, Y.X. and L.W.; Validation, Z.H.; Writing–original draft, Z.H.; Writing–review & editing, Y.Y. All authors have read and agreed to the published version of the manuscript.

Funding: This work was funded by the Ministry of Science and Technology of the People's Republic of China (Granted by No. 2017YFC1601600).

Conflicts of Interest: The authors declare that they have no financial or personal relationships with other people or organizations that could inappropriately influence the work in this study.

References

1. Wang, L.; Zhang, Q.; Yan, Z.; Tan, Y.; Zhu, R.; Yu, D.; Yang, H.; Wu, A. Occurrence and Quantitative Risk Assessment of Twelve Mycotoxins in Eggs and Chicken Tissues in China. *Toxins* **2018**, *10*, 477. [CrossRef]
2. Zhu, R.; Zhao, Z.; Wang, J.; Bai, B.; Wu, A.; Yan, L.; Song, S. A simple sample pretreatment method for multi-mycotoxin determination in eggs by liquid chromatography tandem mass spectrometry. *J. Chromatogr. A* **2015**, *1417*, 1–7. [CrossRef] [PubMed]
3. Wang, Q.; Zhang, Y.; Zheng, N.; Guo, L.; Song, X.; Zhao, S.; Wang, J. Biological System Responses of Dairy Cows to Aflatoxin B1 Exposure Revealed with Metabolomic Changes in Multiple Biofluids. *Toxins* **2019**, *11*, 77. [CrossRef] [PubMed]
4. Iqbal, S.Z.; Nisar, S.; Asi, M.R.; Jinap, S. Natural incidence of aflatoxins, ochratoxin A and zearalenone in chicken meat and eggs. *Food Control* **2014**, *43*, 98–103. [CrossRef]
5. Zhao, Z.; Liu, N.; Yang, L.; Deng, Y.; Wang, J.; Song, S.; Lin, S.; Wu, A.; Zhou, Z.; Hou, J. Multi-mycotoxin analysis of animal feed and animal-derived food using LC–MS/MS system with timed and highly selective reaction monitoring. *Anal. Bioanal. Chem.* **2015**, *407*, 7359–7368. [CrossRef]
6. Mazur-Kuśnirek, M.; Antoszkiewicz, Z.; Lipiński, K.; Fijałkowska, M.; Purwin, C.; Kotlarczyk, S. The effect of polyphenols and vitamin E on the antioxidant status and meat quality of broiler chickens fed diets naturally contaminated with ochratoxin A. *Arch. Anim. Nutr.* **2019**, *73*, 431–444. [CrossRef] [PubMed]
7. Emmanuel, K.T.; Els, V.P.; Bart, H.; Evelyne, D.; Els, D. Carry-over of some Fusarium mycotoxins in tissues and eggs of chickens fed experimentally mycotoxin-contaminated diets. *Food Chem. Toxicol.* **2020**, *145*, 111715. [CrossRef]
8. Buranatragool, K.; Poapolathep, S.; Isariyodom, S.; Imsilp, K.; Klangkaew, N.; Poapolathep, A. Dispositions and tissue residue of zearalenone and its metabolites α-zearalenol and β-zearalenol in broilers. *Toxicol. Rep.* **2015**, *2*, 351–356. [CrossRef]
9. Yan, Z.; Wang, L.; Wang, J.; Tan, Y.; Yu, D.; Chang, X.; Fan, Y.; Zhao, D.; Wang, C.; De Boevre, M.; et al. A QuEChERS-Based Liquid Chromatography-Tandem Mass Spectrometry Method for the Simultaneous Determination of Nine Zearalenone-Like Mycotoxins in Pigs. *Toxins* **2018**, *10*, 129. [CrossRef]
10. Cui, X.; Muhammad, I.; Li, R.; Jin, H.; Guo, Z.; Yang, Y.; Hamid, S.; Li, J.; Cheng, P.; Zhang, X. Development of a UPLC-FLD Method for Detection of Aflatoxin B1 and M1 in Animal Tissue to Study the Effect of Curcumin on Mycotoxin Clearance Rates. *Front. Pharmacol.* **2017**, *8*, 650. [CrossRef]
11. D'Agnello, P.; Vita, V.; Franchino, C.; Urbano, L.; Curiale, A.; Debegnach, F.; Iammarino, M.; Marchesani, G.; Chiaravalle, A.; Pace, R. ELISA and UPLC/FLD as Screening and Confirmatory Techniques for T-2/Ht-2 Mycotoxin Determination in Cereals. *Appl. Sci.* **2021**, *11*, 1688. [CrossRef]
12. McMaster, N.; Acharya, B.; Harich, K.; Grothe, J.; Mehl, H.L.; Schmale, D.G. Quantification of the Mycotoxin Deoxynivalenol (DON) in Sorghum Using GC-MS and a Stable Isotope Dilution Assay (SIDA). *Food Anal. Methods* **2019**, *12*, 2334–2343. [CrossRef]
13. Rodríguez-Carrasco, Y.; Moltó, J.C.; Mañes, J.; Berrada, H. Exposure assessment approach through mycotoxin/creatinine ratio evaluation in urine by GC–MS/MS. *Food Chem. Toxicol.* **2014**, *72*, 69–75. [CrossRef]

14. Turkmen, Z.; Kurada, O. Rapid HPTLC determination of patulin in fruit-based baby food in Turkey. *JPC—J. Planar Chromatogr.-Mod. TLC* **2020**, *33*, 209–217. [CrossRef]
15. Tkaczyk, A.; Jedziniak, P. Development of a multi-mycotoxin LC-MS/MS method for the determination of biomarkers in pig urine. *Mycotoxin Res.* **2021**, *37*, 169–181. [CrossRef] [PubMed]
16. Li, N.; Qiu, J.; Qian, Y. Polyethyleneimine-modified magnetic carbon nanotubes as solid-phase extraction adsorbent for the analysis of multi-class mycotoxins in milk via liquid chromatography–tandem mass spectrometry. *J. Sep. Sci.* **2020**, *44*, 636–644. [CrossRef] [PubMed]
17. Steiner, D.; Malachová, A.; Sulyok, M.; Krska, R. Challenges and future directions in LC-MS-based multiclass method development for the quantification of food contaminants. *Anal. Bioanal. Chem.* **2020**, *413*, 25–34. [CrossRef] [PubMed]
18. den Hollander, D.; Croubels, S.; Lauwers, M.; Caekebeke, N.; Ringenier, M.; De Meyer, F.; Reisinger, N.; Van Immerseel, F.; Dewulf, J.; Antonissen, G. Applied Research Note: Biomonitoring of mycotoxins in blood serum and feed to assess exposure of broiler chickens. *J. Appl. Poult. Res.* **2020**, *30*, 100111. [CrossRef]
19. Nakhjavan, B.; Ahmed, N.S.; Khosravifard, M. Development of an Improved Method of Sample Extraction and Quantitation of Multi-Mycotoxin in Feed by LC-MS/MS. *Toxins* **2020**, *12*, 462. [CrossRef] [PubMed]
20. Castaldo, L.; Graziani, G.; Gaspari, A.; Izzo, L.; Tolosa, J.; Rodríguez-Carrasco, Y.; Ritieni, A. Target Analysis and Retrospective Screening of Multiple Mycotoxins in Pet Food Using UHPLC-Q-Orbitrap HRMS. *Toxins* **2019**, *11*, 434. [CrossRef]
21. Sun, F.; Tan, H.; Li, Y.; De Boevre, M.; Zhang, H.; Zhou, J.; Li, Y.; Yang, S. An integrated data-dependent and data-independent acquisition method for hazardous compounds screening in foods using a single UHPLC-Q-Orbitrap run. *J. Hazard. Mater.* **2020**, *401*, 123266. [CrossRef] [PubMed]
22. Zhou, H.; Cao, Y.-M.; Miao, S.; Lan, L.; Chen, M.; Li, W.-T.; Mao, X.-H.; Ji, S. Qualitative screening and quantitative determination of 569 pesticide residues in honeysuckle using ultrahigh-performance liquid chromatography coupled to quadrupole-Orbitrap high resolution mass spectrometry. *J. Chromatogr. A* **2019**, *1606*, 460374. [CrossRef]
23. Alaboudi, A.R.; Osaili, T.M.; Otoum, G. Quantification of mycotoxin residues in domestic and imported chicken muscle, liver and kidney in Jordan. *Food Control.* **2021**, *132*, 108511. [CrossRef]
24. Zhang, X.; Song, Y.; Jia, Q.; Zhang, L.; Zhang, W.; Mu, P.; Jia, Y.; Qian, Y.; Qiu, J. Simultaneous determination of 58 pesticides and relevant metabolites in eggs with a multi-functional filter by ultra-high performance liquid chromatography-tandem mass spectrometry. *J. Chromatogr. A* **2019**, *1593*, 81–90. [CrossRef] [PubMed]
25. Arce-López, B.; Lizarraga, E.; Flores-Flores, M.; Irigoyen, Á.; González-Peñas, E. Development and validation of a methodology based on Captiva EMR-lipid clean-up and LC-MS/MS analysis for the simultaneous determination of mycotoxins in human plasma. *Talanta* **2020**, *206*, 120193. [CrossRef] [PubMed]

Review

Determination of Antibiotic Residues in Aquaculture Products by Liquid Chromatography Tandem Mass Spectrometry: Recent Trends and Developments from 2010 to 2020

Yueting Xiao [1,2], Shuyu Liu [1,*], Yuan Gao [2,3], Yan Zhang [2], Qinghe Zhang [2,3] and Xiuqin Li [2,3,*]

1. School of Chemistry and Chemical Engineering, Shanghai University of Engineering Science, Shanghai 201620, China; 15797752285@163.com
2. Food Safety Laboratory, Division of Metrology in Chemistry, National Institute of Metrology, Beijing 100029, China; gaoy@nim.ac.cn (Y.G.); yanzhang_larissa@163.com (Y.Z.); zhangqh@nim.ac.cn (Q.Z.)
3. Key Laboratory of Chemical Metrology and Applications on Nutrition and Health for State Market Regulation, Beijing 100029, China
* Correspondence: liushuyu1219@163.com (S.L.); lixq@nim.ac.cn (X.L.); Tel.: +86-10-64524784 (X.L.)

Citation: Xiao, Y.; Liu, S.; Gao, Y.; Zhang, Y.; Zhang, Q.; Li, X. Determination of Antibiotic Residues in Aquaculture Products by Liquid Chromatography Tandem Mass Spectrometry: Recent Trends and Developments from 2010 to 2020. *Separations* **2022**, *9*, 35. https://doi.org/10.3390/separations9020035

Academic Editors: Jongki Hong and Andreas Seubert

Received: 28 December 2021
Accepted: 26 January 2022
Published: 29 January 2022

Publisher's Note: MDPI stays neutral with regard to jurisdictional claims in published maps and institutional affiliations.

Copyright: © 2022 by the authors. Licensee MDPI, Basel, Switzerland. This article is an open access article distributed under the terms and conditions of the Creative Commons Attribution (CC BY) license (https://creativecommons.org/licenses/by/4.0/).

Abstract: The issue of antibiotic residues in aquaculture products has aroused much concern over the last decade. The residues can remain in food and enter the human body through the food chain, posing great risks to public health. For the safety of foods and products, many countries have issued maximum residue limits and banned lists for antibiotics in aquaculture products. Liquid chromatography tandem mass spectrometry (LC/MS/MS) has been widely used for the determination of trace antibiotic residues due to its high sensitivity, selectivity and throughput. However, considering its matrix effects during quantitative measurements, it has high requirements for sample pre-treatment, instrument parameters and quantitative method. This review summarized the application of LC/MS/MS in the detection of antibiotic residues in aquaculture products in the past decade (from 2010 to 2020), including sample pre-treatment techniques such as hydrolysis, derivatization, extraction and purification, mass spectrometry techniques such as triple quadrupole mass spectrometry and high-resolution mass spectrometry as well as status of matrix certified reference materials (CRMs) and matrix effect.

Keywords: aquaculture products; antibiotic residues; liquid chromatography tandem mass spectrometry (LC/MS/MS); sample pre-treatment; matrix effects

1. Introduction

With the rapid development of China's aquaculture industry, China's aquaculture production now accounts for over 60% of the world's total [1]. To achieve the high yield, fish production adopts intensive and semi-intensive practices, which lead to a higher concentration of animals in small spaces and substantially increase the risk of disease [2]. Thus, antibiotics are often used as veterinary drugs and feed additives to treat and prevent aquaculture infections. The misuse or long-term use of antibiotics can lead to resistance in aquaculture products and humans, and even toxic side effects such as teratogenicity, carcinogenicity and mutagenicity in human body [3]. Consequently, many countries have gradually introduced maximum residue limits (MRLs) and prohibition lists for veterinary drugs residues in food of animal origin.

At present, antibiotics commonly used in aquaculture mainly include quinolones (QNs), sulfonamides (SAs), amphenicols (APs), nitrofurans (NFs), tetracyclines (TCs), macrolides (MALs), aminoglycosides (AGs), lincosamides, beta-lactams, etc. In 2002, the use of antibiotics such as chloramphenicol, nitrofuran antibiotics and nitroimidazole in food-producing animals was banned in China. In 2016, the Ministry of Agriculture and Rural Affairs of China announced a ban on the use of four QNs, lomefloxacin, pefloxacin, ofloxacin and norfloxacin in food-producing animals. However, some veterinary drugs that have been banned, such as chloramphenicol, nitrofurans and malachite green, can still be detected in shrimp and fish samples [4]. To further strengthen the control of veterinary drugs, China issued a prohibited list of drugs and other compounds and a standard for maximum residue limits (Supplementary Materials) of veterinary drugs in animal origin food in 2019 (GB 31650-2019). The Codex Alimentarius Commission (CAC) developed the standards of MRLs for veterinary drugs in food (CAC/MRL 2-2015). Additionally, the European Commission (EU) has published the EU No 37/2010 about pharmacologically active substances and their classification regarding MRLs in foodstuffs of animal origin. As permitted veterinary drugs, many antibiotics have available MRL data in the Annex III of EU No 37/2010. For some prohibited antibiotics, the EU had set a minimum required performance level (MRPL) such as nitrofuran metabolites, chloramphenicol and sum of malachite green and leuco-malachite green at 1, 0.3 and 2 $\mu g \cdot kg^{-1}$ in aquaculture products, respectively (Commission Decision 2004/25/EC). Table S1 summarizes the MRL or MRPL of antibiotics in aquaculture products in different countries. The current prohibited antibiotics of aquaculture products in the Chinese standards are basically consistent with those in the EU standards, and both have similar MRLs for most antibiotics. Compared with those in the CAC standard, China's existing veterinary drug residue limits for aquaculture products are more comprehensive. With the improvement of limit standards, the national standard detection method of antibiotics in aquaculture products have also increasingly advanced. There are seventeen relevant standards for antibiotics in aquaculture products in China (Table S2), twelve of which are liquid chromatography tandem mass spectrometry detection methods (LC/MS/MS). According to the above, it can be concluded that LC/MS/MS will be more and more widely used in the detection of antibiotics in aquaculture products.

In 2016, Justino et al. [5] reviewed detection techniques for contaminants in aquaculture products, indicating that LC/MS/MS is becoming the dominant technique. In the same year, Santos et al. [2] summarized the current analytical methods for eight antibiotics in aquaculture fishes, detailing the legal provisions governing antibiotics in different countries and pointing out that multiclass and multiresidue detection is the future trend. In summary, based on the current trends of detection methods, this paper reviews the characteristics and research status of LC/MS/MS for the detection of antibiotics in aquaculture products during last decade (2010–2020), summarizes the sample pre-treatment methods of different antibiotics in aquaculture products, giving emphasis on hydrolysis, derivatization and extraction/purification methods and discusses representative matrix effect of antibiotics. The situation of matrix reference material in different countries is discussed as well.

2. Sample Pre-Treatments

The main steps of the analytical procedures used for determination of multi-antibiotics in Aquaculture products are shown in Figure 1. Aquaculture products are complex foods with high fat and protein, which increases the difficulty of extraction and separation. Thus, prior to analysis, extraction/clean-up and enrichment/concentration techniques are often needed to eliminate or reduce matrix effects to obtain more accurate results. The good chromatography separation and sensitive mass spectrometry response can also effectively improve the accuracy and sensitivity of the analysis. As we can see from Figure 2, most antibiotics are bound to proteins in aquaculture products and require acid hydrolysis prior to extraction, such as NFs, TCs and SAs, among which NFs requires hydrolysis along with derivatization for mass spectrometric detection to improve detection sensitivity. In addition, when using ultraviolet or fluorescence detectors to detect some antibiotics without chromogenic and fluorescent groups, it is also necessary to use derivatization reagents to give the analytes ultraviolet or fluorescent properties, for example, AGs and NFs [6,7]. However, TCs and QNs have chromogenic and fluorescent groups that do not require derivatization. In addition, most antibiotics have high polarity or boiling points and require derivatization before detection by gas chromatography (GC). For example, Santos et al. [8]

used gas chromatography tandem mass spectrometry (GC/MS) to screen chloramphenicol in trout by derivatization with silylated reagents after extraction and purification.

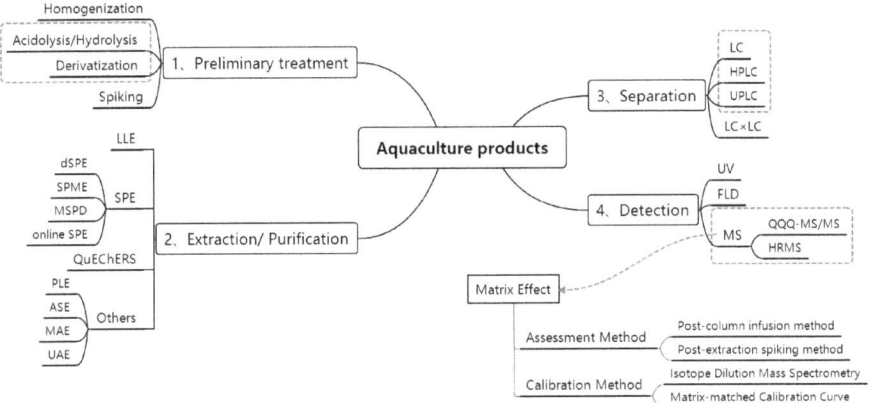

Figure 1. The main steps of the analytical procedure applied in determination of antibiotics of aquaculture products.

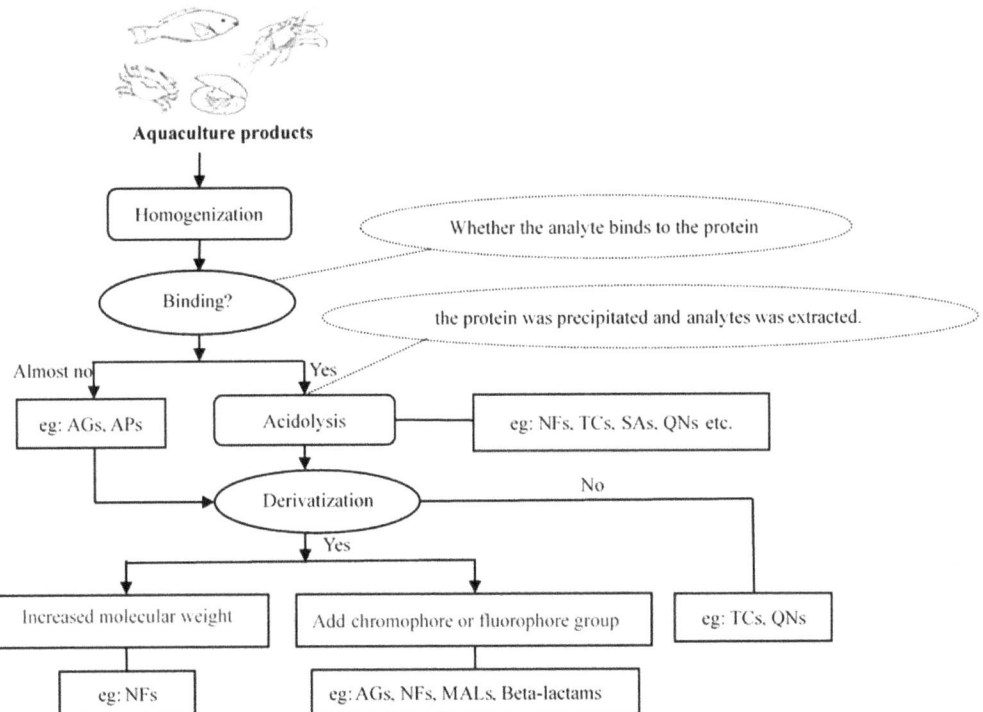

Figure 2. The current workflow of preliminary treatment for major antibiotics.

2.1. Hydrolysis and Derivatization

The hydrolysis step is highly required to convert the combined state to the free state before sample extraction and purification for those antibiotics in aquaculture products in the form of protein binding. In order to provide a theoretical basis for establishing a more efficient pre-treatment method, many researchers further investigated the rules of binding and desorption of proteins and antibiotic drugs [9]. In 2002, M. A. Khan et al. [10] demonstrated the high affinity between bovine serum albumin (BSA) and TCs by fluorescence quenching. In 2018, Pan Lin [9] studied the effect of three different matrix components (protein, fat and water) on the extraction efficiency of TCs, and the results showed that TCs have a strong binding effect with egg albumin (CEA), which can lead to low extraction efficiency. Li et al. [11] reported that there was a strong hydrogen bonding interaction between fluoroquinolone antibiotics (FQs) and fish serum albumin (FSA) which can be broken by 50–90% acetonitrile acid solution, and when protein was precipitated with 90% acetonitrile solution, the recoveries of four FQs were >80%. Zhang Yanxi [12] chose BSA as the model carrier protein to simulate the physiological conditions of fish in vitro, and it was confirmed that sulfamethoxazole and sulfamedoxine interacted with BSA, which lead to the low recovery of SAs. The ammonium acetate buffer including 0.3% acetic acid could effectively eliminate the binding of SAs with BSA, and the recovery reached more than 90%.

In addition, the parent NFs are metabolized rapidly in animals, and the half-lives in vivo are not more than a number of hours [13]. Most of the methods published in the literature rely on the detection of metabolites. Moreover, their metabolites tend to form metabolite-protein adducts that are stable for a long time, so the acidic hydrolysis step is commonly used to liberate the covalently bound metabolites [14]. However, nitrofuran metabolites, such as semicarbazide (SEM), 3-amino-2-oxazolidone (AOZ), 1-amino-hydantoin (AHD) and 5-methylmorpholino-3-amino-2-oxazolidinone (AMOZ), are characterized by small relative molecular mass (75–201 Da) and large polarity, which makes it difficult to detect directly by mass spectrometry. In most articles, free nitrofuran metabolites were derivatized with 2-nitrobenzaldehyde (2-NBA) as the derivatizing reagent under a 37 °C shaking bath for 16 h [15–20] to increase the relative molecular mass and detection sensitivity before extraction. Although hydrolysis and derivatization require a long time, they are the key to an efficient extraction for binding antibiotics. To shorten derivatization time to 2 h, some researchers [21] increased the derivatization temperature to 60 °C in a shaking bath, but the sufficient hydrolysis time of the incurred sample was not discussed in detail. Differently, Tao et al. [22] and Wang et al. [23] adopted an ultrasound-assisted derivatization method to replace the shaking bath method (37 °C, 16 h). With 2-nitrobenzaldehyde (2-NBA) as the derivatization reagent, the NF metabolites were hydrolyzed and derivatized with a reaction temperature of 40 °C for 1 h [22]. Palaniyappan et al. [24] developed a new method of microwave-assisted derivatization, and the results were achieved in a short time of 6 min with good recovery. Moreover, different derivatization reagents have also been proposed. Luo et al. [25] used 7-(diethylamino)-2-oxochromene-3-carbaldehyde (DAOC) as the derivatization reagent to react with four NF metabolites to form hydrazone derivatives under the assistance of a microwave within 20 min, which were very stable and exhibited excellent fluorescence sensitivity with maximum excitation and emission wavelengths of 450 and 510 nm, respectively. Du et al. [6] chose 2-hydroxy-1-naphthaldehyde (HN) as a novel derivatization agent, and the synthetic derivative was easily formed and stable, which was suitable for detection by HPLC-FLD and HPLC-MS/MS.

Other than the use of Nitrofurazone (NFZ), the presence of SEM in the sample may also occur by reaction with biurea and azodicarbonamide that are commonly used for food preservation. In addition, SEM is naturally present in the shells of crayfish, shrimp, prawn and soft-shell crab. Therefore, the use of SEM as the exclusive marker for NFZ might be unreliable. 5-nitro-2-furaldehyde (NF) was used as another residual marker for nitrofurazone, and 2,4-dinitrophenylhydrazine (DNPH) was used as a derivatization reagent [26,27]. Derivatization was easily performed in an ultrasonic water bath at 30 °C for 5 min [26], greatly shortening the derivatization time.

2.2. Extraction and Purification Methods

Extraction and purification methods for antibiotics in aquaculture products mainly include liquid–liquid extraction (LLE), solid-phase extraction (SPE), QuEChERS, pressurized liquid-phase extraction (PLE), microwave-assisted extraction (MAE), etc.

2.2.1. Liquid–Liquid Extraction (LLE)

LLE was traditionally used for the extraction of antibiotics due to its simplicity and practicality. For LLE, solvent selection plays a critical role to enhance the recovery of the analyte, which improves the limit of detection (LOD), and minimize the matrix effect. The extraction solvent was chosen according to the physicochemical properties of compounds. Du et al. [6] and Zhang et al. [28] chose ethyl acetate as the extract solvent to extract four nitrofuran compounds in shrimp, with a recovery rate of 85–107%. As for AGs with greater polarity, a simple extraction is generally performed with an aqueous buffer [29,30]. Kaufmann et al. [29] used a trichloroacetic acid aqueous solution for extraction, followed by solid-phase extraction, with recoveries of 60–85%. Additionally, different acids, bases, salts or complexing agents are usually added to improve the extraction efficiency and ionization efficiency of the analytes. Manuel et al. [31] adopted an acetonitrile solvent with 5% formic acid to extract eight quinolones in fish samples; the formic acid provided an acidic medium to facilitate the extraction of quinolone antibiotics with a recovery rate of 72–108% and intra-day reproducibility of less than 10.5%. The mixed solution of ethyl acetate and ammonia water (98:2) were used as the extraction solvent to extract three APs in tilapia; the ammonia water played a role in facilitating the extraction, and the recovery rate was 79.8–92.0% [32]. As for TCs, QNs or gentamicin, which are easily complexed with polyvalent metal cations, chelating agents are often added to the extraction solvent, then, sodium sulfate is often used in the phase separation step instead of magnesium sulfate [33–36]. Grande-Martinez et al. [37] developed a modified QuEChERS procedure to extract TCs in fish. A fish sample was extracted twice by an EDTA-McIlvaine buffer and acetonitrile, then 50 mg of C18 was added for further purification, with recoveries ranging from 80 to 105%. Shin et al. [38] performed a two-step solvent extraction method. The aqueous phase extraction solution was added with Ethylene Diamine Tetraacetic Acid (EDTA) and ammonium acetate, and acetonitrile was added with ammonium formate. The results showed that the addition of EDTA can increase the extraction recovery rate of tetracycline from less than 60% to nearly 100%. Because of the high fat in aquaculture products, the additional step of degreasing is necessary. N-hexane is the most common degreasing solvent [39–41]. Meanwhile, it is recommended to increase the centrifugation speed or add sodium chloride to overcome emulsification [38].

For the extraction of antibiotics with similar polarity, most extraction methods use single organic solvents as extractants, however, when applied to multiresidue antibiotics with different physical and chemical properties, water or a water buffer can be combined with organic solvents to expand the extraction range of analytes [39,40]. Jia et al. [42] developed a multiresidue method for the analyses of 137 veterinary drug residues. Extraction of compounds was achieved by 5 mL of an acetonitrile/water solution (84/16, v/v) with one hundred microliters of 0.1 M of EDTA and 1% acetic acid, and then Primary-secondary amine (PSA) and Z-Sep+ as the adsorbent of solid-phase microextraction (SPME) to purify, with a good recovery rate ranging from 82 to 112%. Figure 3 summarizes the commonly used extraction solvents for different antibiotics, including acetonitrile, methanol, ethyl acetate, buffer solutions and so on. For nitrofuran antibiotics, most of articles used ethyl acetate as the extraction solvent, and sometimes it is mixed with a small proportion of acetonitrile for extraction. For the more polar aminoglycosides, different buffer solutions were the main extraction solvents. Acidified acetonitrile is commonly used for the simultaneous extraction of quinolones, sulfonamides and tetracyclines.

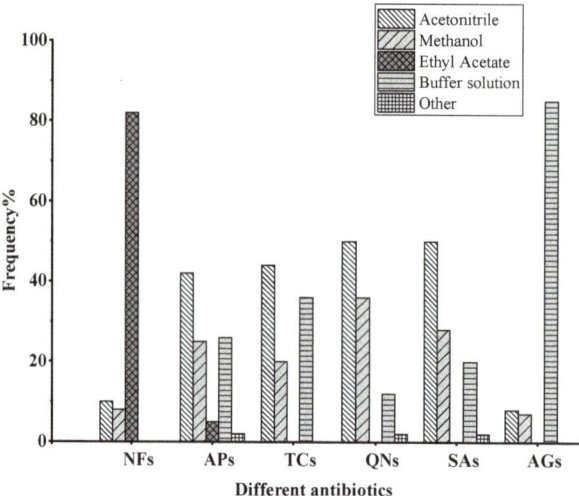

Figure 3. The proportion of different extraction solvents used for five kinds of antibiotics in articles from the recent decade.

2.2.2. Solid-Phase Extraction (SPE)

C18 [43,44], C8, Phenyl [45] and HLB [28,46] are the main reverse-phase sorbent materials used for solid-phase extraction and purification of antibiotics from aquaculture products. Furthermore, Oasis HLB SPE column is more common in the extraction of antibiotics from aquaculture products [14,22,47–49]. Evaggelopoulou et al. [46] proposed an approach to extract six penicillin antibiotics and three APs from gilthead seabream tissues. The extraction was carried out by a mixture of H_2O/acetone (50/50% v/v), which was repeated twice in order to increase the rates of recovery. Subsequently, the recoveries of Lichrolut RP-18 and OASIS HLB SPE column were compared. The results showed that the recovery rate of the OASIS HLB SPE column was higher, which could reach more than 95%. Liu et al. [47] used Ultrasonic-Assisted Extraction (UAE) combined the SPE method to determine the multiresidue antibiotics in fish and plasma. Fish samples were extracted with methanol and enriched using Oasis HLB solid-phase extraction columns in one step, the average recovery was 61–111%, and relative standard deviation (RSD) was less than 25%. For highly polar aminoglycoside antibiotics, ion exchange extraction columns are more suitable [50]. Gbylik et al. [51] developed a two-step extraction mean to separate seven classes of antibiotics. including aminoglycosides from fish. Firstly, the isolation of residues from the sample was applied by m-phosphoric acid and heptafluorobutyric acid as an ion-pair agent and acetonitrile, then, a clean-up technique was performed by polymeric weak cationic extraction column (Strata X-CW), the result of recovery was from 96 to 111%.

In recent years, increasingly new solid-phase extraction sorbents have been applied, such as graphene, multiwalled carbon nanotubes (MWCNTs), molecularly imprinted polymers (MIPs) [52–54]. Wu et al. [35] proposed two-dimensional (2D) planar graphene powder as an SPE sorbent for enrichment and cleanup of MLs from a carp sample. Finally, 15 mg of graphene was selected when the carp sample was 1 g, and the extraction recoveries ranged from 81.7 to 110.5%. With the development of analytical techniques, some new techniques based on traditional SPE have been applied in the detection of antibiotic residues in aquaculture products, such as solid-phase microextraction (SPME), matrix solid-phase dispersion extraction (MSPD), and dispersive solid-phase extraction (d-SPE). Mondal et al. [55] synthesized a novel MIL-101(Cr)-NH_2 fiber for extraction of six antibiotics (flumequine, Nalidixic acid, tilmicosin, sulfadimethoxine, sulfaphenazole and methomyl pyrimidine) from fish meat by SPME, with better reproducibility than conventional fibers, and the precision is between 1.5 and 8.3%. Pan et al. [52] extracted

AGs from fish by MSPD and compared the extraction efficiency of two sorbents, C18 and graphitized carbon black (GCB), indicating that the recoveries with the use of C18 were higher than those with GCB. Shen et al. [53] proposed a micropipette-matrix solid-phase dispersion (PT-MSPD) technique treated with a pipette tip and dispersant HLB for the detection of 14 QNs in fish tissues, and the absolute recoveries were 25% higher than those of conventional MSPD. Unlike traditional SPE, MSPD does not require tissue homogenization, precipitation, centrifugation, pH adjustment and sample transfer, avoiding the loss of samples, shortening the operation time and saving organic reagents. D-SPE is a technique in which the solid-phase extraction sorbent is dispersed in the extraction solution of the sample [56,57] and is also commonly used in the QuEChERS method [34,42], especially for the extraction of antibiotics from complex matrices. For example, Manuel et al. [31] extracted eight quinolone antibiotics from a variety of complex fish matrices by simple acidified acetonitrile liquid–liquid extraction, and then d-SPE was performed by C18 and $MgSO_4$. The recoveries were from 72 to 108% with good reproducibility and RSD less than 6.4%.

2.2.3. Other Techniques

To enhance environmental protection, some green extraction techniques have been gradually applied, such as pressurized liquid extraction (PLE) or accelerated solvent extraction (ASE), microwave-assisted extraction (MAE), ultrasound-assisted extraction (UAE), etc. Compared with traditional methods, these extraction techniques take advantage of saving extraction times and reducing solvent consumption. Liu et al. [58] adopted the PLE method to extract TCs in fish and shrimp. A mixed solvent of trichloroacetic acid (TCA)/methanol (1:3) was the solvent. Equal amounts of Na2EDTA should be added before PLE. The recoveries ranged from 75.6 to 103.5%. PLE reduced the use of solvents and extraction time compared to traditional liquid–liquid extractions. Hoff et al. [59] simultaneously detected 16 SAs in liver, comparing two extraction methods of PLE and UAE, with recoveries close to 100%, but the latter with a slight advantage of UAE in terms of solvent usage and time required. Kazakova et al. [60] extracted multiple antibiotics from lobsters by LLE and MAE methods. The best condition of MAE was 50 µL Proteinase-K and 5 µL formic acid (FA) at 50 W for about 5 min. The recoveries ranged from 71 to 100%.

In summary, LLE is a traditional and easy-to-operate but time-consuming method that is more suitable for the extraction of antibiotics with similar polarity. Compared to LLE, SPE is more automated and enables simultaneous extraction and enrichment. With the development of science and technology, some advanced materials, such as MIP materials, magnetic materials and so on, provide SPE with excellent performance. At present, SPE and LLE are still important methods for extracting and purifying antibiotics in aquaculture products. However, in our prospect, microextraction techniques and green extraction techniques are bound to become the mainstream in the near future.

3. Liquid Chromatography-Mass Spectrometry Detection Technique

The common ion sources for the liquid chromatography tandem mass spectrometry method are electrospray ionization (ESI) and atmospheric pressure chemical ionization (APCI). ESI sources are mainly used for polar and macromolecular compounds and have a wider range of application, so most antibiotic detection is often performed using ESI source mass spectrometry. Table 1 outlines applications of LC/MS/MS for the analysis of antibiotics in aquaculture products during the recent decade. Aldeek et al. [16] used LC-ESI-MS/MS to simultaneously detect four nitrofuran metabolites and chloramphenicol in tilapia and shrimp, and quantified using an isotopic internal standard, which could calibrate the loss of analytes during sample preparation well. The recovery rate reached 90–100%, which was 40% higher than that of the external standard method, and the RSD was less than 10%. Kung et al. [56] adopted the QuEChERS method for the detection of four sulfonamides in fish meat by HPLC-ESI-MS/MS. The recoveries were 80.2–93.5% with RSD less than 9%, the decision limit (CCα) ranged from 1.49 to 10.09 $µg·kg^{-1}$, and detection capability (CCβ) ranged from 1.71 to 11.4 µg·kg^{-1}. Jansomboon et al. [61] detected four sulfonamides in fish by LC-ESI-MS/MS after acidic methanol and acetonitrile liquid–liquid extraction, the detection limit was from 0.75 to 3.13 µg·kg^{-1}. However, previous studies have demonstrated that the APCI source had better sensitivity for SEM. An et al. [15] found that that the detection signal generated by the APCI source was from three to fivefold higher than that of the ESI source. The APCI source also had lower background noise, which significantly enhanced the SEM signal. The LOD was 0.052–0.108 µg·kg^{-1}, the LOQ was 0.25 µg·kg^{-1}, and the recovery was 100.2–104.0% with good reproducibility. Similarly, Chumanee et al. [62] chose the APCI source rather than the ESI source for the detection of SEM in order to improve the sensitivity, with LODs of 0.1–0.3 µg·kg^{-1} and LOQs of 0.1–0.5 µg·kg^{-1}.

With the increasing development of mass spectrometry, there is an increasing number of multiclass and multiresidue analysis (MCMR) methods for the simultaneous screening or quantification of dozens or even hundreds of different classes of residues in samples [63–67]. A subject search of the ScienceDirect database using keywords related to LC-MS, multiple residues, antibiotics and aquaculture product indicated an overall growth trend for the last 10 years (Figure 4a). Miossec et al. [68] established a UPLC-QQQ-MS/MS method for the simultaneous detection of 42 veterinary drugs in four kinds of seafood with a simple LLE using acidified methanol extraction, followed by enrichment and filtration, and the LODs ranged from 0.1 to 5.0 µg·kg^{-1} for all antibiotics except amoxicillin. Dasenaki et al. [63] used a UPLC-QQQ-MS/MS method to simultaneously detect up to 20 categories of 115 veterinary drugs, and the recoveries of 80% of the analytes ranged from 50 to 120% with RSD less than 18%. The LOQs for all analytes were less than 5 µg·kg^{-1} except for dalfloxacin, of which it was 5.6 µg·kg^{-1}. Jia et al. [42] developed a UHPLC-Q/Obitrap-HRMS method for simultaneous analysis of one hundred and thirty-seven veterinary drug residues and metabolites from sixteen different classes in tilapia. Three ways of data acquisition were compared: Fall Scan/dd-MS/MS, Fall Scan/all-ion fragmentation (AIF) and Fall Scan/variable data independent acquisition (vDIA). The result showed that using vDIA instead of dd-MS/MS or AIF for nontargeted generation of fragment ions improved the selectivity and sensitivity of the analysis. The recoveries of 137 analytes ranged from 81 to 111%, CCα ranged from 0.01 to 2.73 µg·kg^{-1}, and CCβ ranged from 0.01 to 4.73 µg·kg^{-1}. Munaretto et al. [57] used LC-Q/TOF-HRMS to detect 182 pesticides, veterinary drugs and other contaminant residues and evaluated the effect of two different scanning methods (FS and dd-MS/MS). It turns out that the FS mode could detect 84% of the compounds, while dd-MS/MS scan could only detect 72%, but dd-MS/MS scan could provide fragmentation information of the target, therefore using dd-MS/MS scan for characterization and the FS mode for quantification.

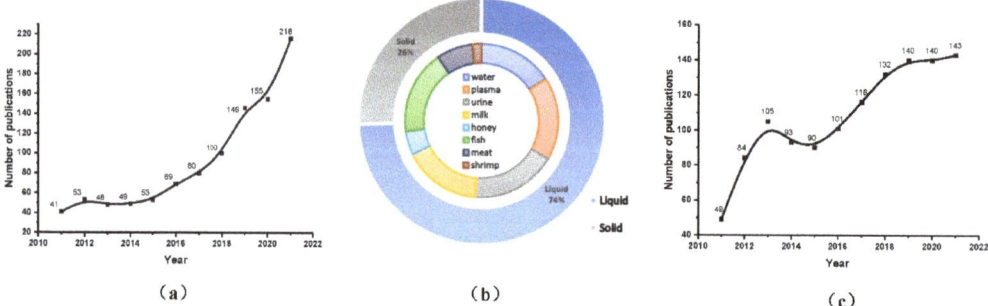

Figure 4. (a) The trend of multiresidue detection of antibiotics in last decade. (b) The application and trend of on-line solid-phase extraction in different matrixes. (c) The trend of using the on-line SPE-LC-MS method in detection of antibiotics in last decade.

Table 1. The applications of LC/MS for the analysis of antibiotics in aquaculture products during last decade.

Analyte (Number)	Preliminary Treatment	Extraction and Purification	Recovery	Detection Method	LODs/CCα (μg·kg^{-1})	LOQs/CCβ (μg·kg^{-1})	Ref.
NFs (4)		SPE: HLB	88~112%	UPLC-ESI-MS/MS	0.5	1.5	[14]
NFs (4)		EtOAc; Hex	101.6~105.9%	LC-APCI-MS/MS	0.05~0.2	0.25	[15]
NFs (4), APs (2)		H$_2$O; EtOAc	85~110% (APs)	UPLC-ESI-MS/MS	0.1~1	0.25~1	[16]
NFs (8), CAP	Acidolysis (HCl), derivatization (2-NBA, shock, 37 °C, 16 h)	EtOAc; Hex, SPE by HLB	97~108% (expect PSH, DNSH, NPIR)	UHPLC-ESI-HRMS	0.01~0.1 [a]	0.01~0.18 [b]	[18]
NFs (4)		EtOAc; Hex	84~115%	LC-ESI-MS/MS	0.1~0.8 [a]	0.3~0.9 [b]	[19]
NFs (8)		EtOAc; Hex	91.6~107.3%	UHPLC-ESI-MS/MS	0.01~0.2; NSTY (2.0)	0.04~0.5; NSTY (5.0)	[41]
NFs (4)	Acidolysis (TCA), derivatization (2-NBA, ultrasound, 40 °C, 1 h)	ASE: MEOH/5% TCA (1/1, v/v); SPE: HLB	77.2~97.4%	LC-ESI-MS/MS	0.07~0.13 [a]	0.31~0.49 [b]	[22]
SEM	Acidolysis (HCl), derivatization (DNPH, ultrasound, 30 °C, 5 min)	EtOAc; SPE: neutral alumina and HLB	80.8~104.4%	UPLC-ESI-MS/MS	0.05	0.1	[26]
APs (3)	/	ACN/H$_2$O (1:1, v/v), EtOAc; MSPD: C18	82.4~99.8%	UPLC-ESI-MS/MS	0.02~0.06 [a]	0.11~0.16 [b]	[52]
APs (3)	/	EtOAC:NH$_3$:H$_2$O (98:2, $v:v$)	80~92%	HPLC-ESI-MS/MS	0.019~54.9 [a]	0.068~64.88 [b]	[32]
FF, FFA	/	EtOAC:NH$_3$:H$_2$O (98:2, $v:v$); Hex, SPE: Phenyl	96.3%(FF),83.0%(FFA)	UPLC-ESI-MS/MS	6 (FF) 1 (FFA)	25 (FF) 3 (FFA)	[45]
SAs (18)	Acidolysis	UAE: 0.1% FA in ACN; On-line SPE: MCX and HLB	71.5~102%	Online-SPE-UHPLC—MS/MS	1.46-15.5	4.90-51.6	[48]
SAs (16)	Acidolysis	ASE: 0.2% FA in ACN; Hex	about 100%	HPLC-QqLIT-MS/MS	10	25	[59]

Table 1. Cont.

Analyte (Number)	Preliminary Treatment	Extraction and Purification	Recovery	Detection Method	LODs/CCα (μg·kg^{-1})	LOQs/CCβ (μg·kg^{-1})	Ref.
SAs (14)	/	ACN; d-SPE: C18	80.2~93.5%	HPLC-ESI-MS/MS	1.49–10.9 [a]	1.71–11.4 [b]	[56]
SAs (14)	/	PT-MSPD: HLB	83~90%	UPLC-ESI-MS/MS	2.3–16.4	6.9–54.7	[53]
TCs (5)	/	EDTA-McIlvaine, ACN; SPE: C18	80~105%	LC-ESI-MS/MS	0.5–1.3	1.7–4.4	[37]
AGs (13)	/	0.05 g/mL TCA in water; SPE: MCX	45–85%	UHPLC-ESI-MS/MS	/	2–25	[29]
AGs (9)	/	USA: PBS; SPE: Poly-Sery and MCX	65~110%	HPLC-ESI-MS/MS	2–10 [a]	6–25 [b]	[30]
MAs (5)	/	MeOH; SPE: Graphene; Hex	81.7~110.5%	UPLC-Qtrap-MS/MS	0.09–0.72	0.3–0.77	[35]
MAs (6)	/	MeOH; Hex	77~109%	HPLC-QTOF-HRMS	5.8–27	17–82	[36]
QNs, SAs, TCs	Acidolysis	1% FA in water, ACN	83–100%	UHPLC-ESI-MS/MS	107–114 [a]	112–129 [b]	[40]
NFs, NIIMs, CAP, MG	Acidolysis (HCl), derivation (2-NBA, 60 °C, 2 h)	EtOAC, ACN, Hex, MgSO$_4$	77.2~125.6%	HPLC-ESI-MS/MS	0.07–1.65 [a]	/	[21]
TCs, APs, SAs, TMP, FQNs, MALs	/	MeOH; SPE: HLB	61~111%	UHPLC-ESI-MS/MS	/	0.03–6.67	[49]

/: not mentioned in the article; EtOAc: ethyl acetate; MEOH: Methanol; ACN: acetonitrile; Hex: n-Hexane; FA: formic acid; TCA: trichloroacetic acid; PBS: phosphate buffered saline; LLE: liquid-liquid extraction; SPE: solid-liquid extraction; d-SPE: dispersive solid-phase extraction; MSPD: Matrix solid-phase dispersion; ASE: accelerated solvent extraction; UAE: ultrasonic assisted extraction; NFs: nitrofurans; APs: amphenicols; SAs: sulfonamides; MALs: macrolides; TCs: tetracyclines; AGs: aminoglycosides; NIIMs: nitroimidazoles; CAP: chloramphenicol; FF: florfenicol; FFA: florfenicol amine; SEM: semicarbazide; TMP: trimethoprim; MG: malachite green; HRMS: high resolution mass spectrometer; APCI: atmospheric pressure chemical ionization; ESI: electrospray ionization; HLB: hydrophilic lipophilic balance; LOD: limit of detection; LOQ: limit of quantitation. [a] using CCα to replace LODs; [b] using CCβ to replace LOQ.

In order to improve the analysis efficiency and automation level, many instruments have realized the coupling of on-line solid-phase extraction and liquid mass spectrometry, like the solid-phase extraction column and the chromatographic column combined through a valve [69]. The target is directly eluted from the solid-phase extraction column to the chromatographic column, which not only simplifies the experimental steps and avoids the loss of analytes, but also greatly improves the sensitivity of the analytical method. A subject search of the ScienceDirect database using keywords related to on-line SPE-LC-MS and antibiotics indicated an overall growth trend for the last 10 years (Figure 4c). However, this method is still mostly used for liquid samples such as environmental water, plasma and urine, while the application of solid samples is less frequent, accounting for only about 26% (Figure 4b). Ma et al. [70] performed rapid determination of 15 sulfonamide antibiotic residues in pork and fish by Online SPE-LC-MSMS with simple extraction using acetonitrile solution containing 2% formic acid and cleanup using the Oasis HLB on-line column (10 mm × 1 mm, Waters, Milford, MA, USA). The recoveries ranged from 78.3 to 99%, RSD was less than 10%, and the LOQs were found to be 0.25–5 $\mu g \cdot kg^{-1}$. Hurtado et al. [71] achieved on-line analysis of 13 analytes including sulfonamides and tetracyclines in catfish. Three kinds of on-line solid-phase extraction columns (C8, C18, GP) are compared. Among them, the recovery of GP was the best, reaching 80% to 99%. The LODs and LOQs were found to be less than 0.1 $\mu g \cdot kg^{-1}$ and 2.4 $\mu g \cdot kg^{-1}$, respectively. On-line SPE can achieve extraction and purification better, reduce matrix effects by diluting samples, avoid loss of analytes and improve sensitivity, thus it has become a hotspot of analytical work. However, the practical application of on-line SPE still has certain limitations due to the few types and high prices of instruments and columns.

4. Matrix Effect

The matrix effect (ME) is a prevalent phenomenon in mass spectrometry analysis, manifested as signal suppression or enhancement, which can affect method sensitivity, precision and accuracy. In recent years, an increasing amount of research about LC-MS/MS method has evaluated the impact of the matrix effect on detection and proposed solutions to reduce or eliminate it. The method of qualitative evaluation of the matrix effect is the post-column infusion method [72], which is assessed by observing the variation in the ESI response of the injected analyte. A constant amount of standard solution of analyte is delivered by an infusion pump. A blank sample extract is injected on the LC column. Then, both of them are mixed through a straight tee into the ion source of the mass spectrometer. Finally, it is easy to identify chromatographic regions most likely to experience matrix effects [73]. The method of quantitative evaluation of the matrix effect is the post-extraction spiking method, which is used to compare the response of the pure solution calibration solution (A) with the matrix-matching standard solution (B) of the same concentration. ME can be measured using the following equation: ME% = B/A × 100.

Grande-Martínez et al. [37] evaluated the ME of five TCs in salmon using the post-extraction spiking method and reduced ME by optimizing sample preparation methods. Due to the high fat content of the salmon matrix, the ME of TCs ranged from 61 to 140%. Then, the author adopted d-SPE with Z-Sep+ to remove matrix interference, and the ME reduced to 95–105% and was almost negligible. Grabicova et al. [65] also used the post-extraction spiking method to assess the matrix effects of 74 drugs in five different fish tissues (liver, kidney, brain, muscle and plasma), and the results showed that the ME was various in different tissues. Tissues with higher lipid content (liver, kidney and brain) were more affected by the matrix, suppressing 50–60% of the response signal. Signal enhancement occurred mostly in muscle and plasma, and matrix signal suppression effects were mostly seen in other tissues. Matrix-matched calibration solution calibration factors were used to calculate analyte concentrations in Grabicova's study. Miossec et al. [68] evaluated the ME of 42 veterinary drugs in four matrices (cod, red mullet, flounder and shrimp). The majority of compounds have matrix suppression effects, while erythromycin A has greater matrix enhancement. Different compounds also have certain differences in different matrices, red mullet has a stronger matrix suppression effect than the other three. Since not every compound had access to its suitable isotopic internal standard, the matrix-matched calibration method was used to compensate for ME. Kim et al. [74] optimized the chromatographic separation gradient and used the isotope internal standard calibration method to compensate the influence of ME of four nitrofuran metabolites. In summary, ME seriously affects the analytical results of LC/MS/MS methods, thus, it is critical to overcome or reduce the effect of ME as much as possible. There are some useful methods to reduce or eliminate ME, including optimizing the sample preparation method [37,75], chromatographic mass spectrometry conditions and parameters [76] and the quantitative calibration method [65,68], especially the isotope dilution mass spectrometry method (IDMS) calibration method [74,77]. With the increasing use of mass spectrometry in the field of analytical chemistry, IDMS is playing an increasingly important role due to its greater accuracy than other calibration methods and its ability to compensate for matrix effects. However, the IDMS method also has some disadvantages, including the cost and availability of suitable isotopic materials, and differences in the physical and chemical properties between the analyte and the isotopic analogue, which can affect the ions generated in the mass spectrometer. Optimizing the sample preparation method is also the most effective way to reduce or eliminate ME, because this method could essentially reduce the matrix in the sample.

5. Antibiotic Food Matrix CRMs

Animal-derived foods such as aquaculture products have high protein and fat content, which have complex matrix interference, thus, matrix-certified reference materials (CRMs) are important for quality control in daily laboratory testing. Up to 2021, China has released a total of 22 matrix CRMs for antibiotic residue analysis (Table S3), involving five types of matrices: fish, honey, chicken, milk powder and egg. The target substances include: nitrofuran metabolites, quinolones, amphenicol, sulfonamide, nitroimidazole. Other countries have also released some related matrix CRMs, for instance, a nitrofuran marker residue in freeze-dried shrimp (MX012A, MXB12B) issued by the National Metrology Institute of Australia (NMIA, Canberra in Australia). The Korea Research Institute of Standards and Science (KRISS) released CRMs of enrofloxacin residues in chicken meal (108-03-003 (130708)) and ciprofloxacin residues in chicken meal (108-03-004 (130715)). The National Research Council Canada (NRC, Ottawa in Canada) issued CRM of veterinary drug residue in bovine (A33-11-02-BOTS). In summary, there are relatively many studies on antibiotic matrix reference materials in China, covering typical matrixes and target substances. However, the current quantity is far from meeting the quality control requirements of antibiotics in aquaculture products.

6. Conclusions

Advanced chemical analysis technology is essential for the development of food analysis. At present, LC/MS/MS technology is widely used for antibiotic detection in aquaculture products. Meanwhile, the trend is shifting towards multiresidue and multiclass detection. Considering the different properties of antibiotics, a suitable pre-treatment method is the key to improving the detection limit of the high-throughput analysis method. The ME should be observed when MS is used. Therefore, the preparation of new materials for sample pre-treatment, the assessment and elimination of matrix effects, the development of matrix CRMs and the combined use of on-line solid-phase extraction and liquid chromatography mass spectrometry are still the main hotspots for trace detection of antibiotics in aquaculture products.

Supplementary Materials: The following supporting information can be downloaded at: https://www.mdpi.com/article/10.3390/separations9020035/s1, Table S1: Maximum Residue Limits ($\mu g \cdot kg^{-1}$) of antibiotics in aquaculture products in various countries, Table S2: Standard methods for the detection of antibiotics in aquaculture products in China, Table S3: Matrix certified reference materials for antibiotic in different countries.

Author Contributions: Conceptualization, data curation, visualization, writing—original draft preparation, Y.X.; writing—review and supervision, S.L., Y.G. and Y.Z.; conceptualization, writing—review and editing and supervision, X.L. and Q.Z. All authors have read and agreed to the published version of the manuscript.

Funding: This research was funded by the Research and Application of the Common Technology of National Quality Infrastructure (funding number: 2016YFF0201106).

Institutional Review Board Statement: Not applicable.

Informed Consent Statement: Not applicable.

Data Availability Statement: Not applicable.

Conflicts of Interest: The authors declare no conflict of interests.

References

1. Mo, W.Y.; Chen, Z.; Leung, H.M.; Leung, A.O.W. Application of veterinary antibiotics in China's aquaculture industry and their potential human health risks. *Environ. Sci. Pollut. Res.* **2015**, *24*, 8978–8989. [CrossRef] [PubMed]
2. Santos, L.; Ramos, F. Analytical strategies for the detection and quantification of antibiotic residues in aquaculture fishes: A review. *Trends Food Sci. Technol.* **2016**, *52*, 16–30. [CrossRef]
3. Sun, M.; Chang, Z.; Van den Brink, P.J.; Li, J.; Zhao, F.; Rico, A. Environmental and human health risks of antimicrobials used in Fenneropenaeus chinensis aquaculture production in China. *Environ. Sci. Pollut. Res. Int.* **2016**, *23*, 15689–15702. [CrossRef] [PubMed]
4. Liu, X.; Steele, J.C.; Meng, X.Z. Usage, residue, and human health risk of antibiotics in Chinese aquaculture: A review. *Environ. Pollut.* **2017**, *223*, 161–169. [CrossRef]
5. Justino, C.I.L.; Duarte, K.R.; Freitas, A.C.; Panteleitchouk, T.S.L.; Duarte, A.C.; Rocha-Santos, T.A.P. Contaminants in aquaculture: Overview of analytical techniques for their determination. *TrAC* **2016**, *80*, 293–310. [CrossRef]
6. Du, N.N.; Chen, M.M.; Sheng, L.Q.; Chen, S.S.; Xu, H.J.; Liu, Z.D.; Song, C.F.; Qiao, R. Determination of nitrofuran metabolites in shrimp by high performance liquid chromatography with fluorescence detection and liquid chromatography-tandem mass spectrometry using a new derivatization reagent. *J. Chromatogr. A* **2014**, *1327*, 90–96. [CrossRef]
7. Luo, X.; Yu, Y.; Kong, X.; Wang, X.; Ji, Z.; Sun, Z.; You, J. Rapid microwave assisted derivatization of nitrofuran metabolites for analysis in shrimp by high performance liquid chromatography-fluorescence detector. *Microchem. J.* **2019**, *150*. [CrossRef]
8. Santos, L.; Barbosa, J.; Castilho, M.C.; Ramos, F.; Ribeiro, C.A.F.; Silveira, M.I.N.d. Determination of chloramphenicol residues in rainbow trouts by gas chromatography–mass spectometry and liquid chromatography–tandem mass spectrometry. *Anal. Chim. Acta* **2005**, *529*, 249–256. [CrossRef]
9. Pan, L. Studies on the Binding Effection of Protein with Pesticides or Veterinary Drugs and Their Detective Application. Master's Thesis, Yantai University, Yantai, China, 2018.
10. Mateen, A.; Khan, S.M.; Musarrat, J. Differential binding of tetracyclines with serum albumin and induced structural alterations in drug-bound protein. *Int. J. Biol. Macromol.* **2002**, *30*, 243–249. [CrossRef]
11. Li, H.; Yin, J.; Liu, Y.; Shang, J. Effect of protein on the detection of fluoroquinolone residues in fish meat. *J. Agric. Food Chem.* **2012**, *60*, 1722–1727. [CrossRef]
12. Zhang, Y. The Interaction of Antibiotics with Protein and High Throughput Screening of Drug Residues in Fish Based on HPLC-Q-TOF-MS. Master's Thesis, Shaanxi University of Science and Technology, Xi'an, China, 2019.
13. Leitner, A.Z.P.; Lindner, W. Determination of the metabolites of nitrofuran antibiotics in animal tissue by high-performance liquid chromatography–tandem mass spectrometry. *J. Chromatogr. A* **2001**, *939*, 49–58. [CrossRef]
14. Zhang, Y.; Qiao, H.; Chen, C.; Wang, Z.; Xia, X. Determination of nitrofurans metabolites residues in aquatic products by ultra-performance liquid chromatography-tandem mass spectrometry. *Food Chem.* **2016**, *192*, 612–617. [CrossRef] [PubMed]
15. An, H.; Henry, M.; Cain, T.; Tran, B.; Paek, H.C.; Farley, D. Determination of Total Nitrofuran Metabolites in Shrimp Muscle Using Liquid Chromatography/Tandem Mass Spectrometry in the Atmospheric Pressure Chemical Ionization Mode. *J. AOAC Int.* **2012**, *95*, 1222–1233. [CrossRef] [PubMed]
16. Aldeek, F.; Hsieh, K.C.; Ugochukwu, O.N.; Gerard, G.; Hammack, W. Accurate Quantitation and Analysis of Nitrofuran Metabolites, Chloramphenicol, and Florfenicol in Seafood by Ultrahigh-Performance Liquid Chromatography–Tandem Mass Spectrometry: Method Validation and Regulatory Samples. *J. Agric. Food. Chem.* **2017**, *66*, 5018–5030. [CrossRef] [PubMed]
17. Fernando, F.; Munasinghe, D.M.S.; Gunasena, A.R.C.; Abeynayake, P. Determination of nitrofuran metabolites in shrimp muscle by liquid chromatography-photo diode array detection. *Food Control.* **2017**, *72*, 300–305. [CrossRef]
18. Kaufmann, A.; Butcher, P.; Maden, K.; Walker, S.; Widmer, M. Determination of nitrofuran and chloramphenicol residues by high resolution mass spectrometry versus tandem quadrupole mass spectrometry. *Anal. Chim. Acta* **2015**, *862*, 41–52. [CrossRef] [PubMed]

19. Øye, B.E.; Couillard, F.D.S.V. Complete validation according to current international criteria of a confirmatory quantitative method for the determination of nitrofuran metabolites in seafood by liquid chromatography isotope dilution tandem mass spectrometry. *Food Chem.* **2019**, *300*, 125175. [CrossRef]
20. Valera-Tarifa, N.M.; Plaza-Bolaños, P.; Romero-González, R.; Martínez-Vidal, J.L.; Garrido-Frenich, A. Determination of nitrofuran metabolites in seafood by ultra high performance liquid chromatography coupled to triple quadrupole tandem mass spectrometry. *J. Food Compos. Anal.* **2013**, *30*, 86–93. [CrossRef]
21. Chen, D.; Delmas, J.M.; Hurtaud-Pessel, D.; Verdon, E. Development of a multi-class method to determine nitroimidazoles, nitrofurans, pharmacologically active dyes and chloramphenicol in aquaculture products by liquid chromatography-tandem mass spectrometry. *Food Chem.* **2020**, *311*, 125924. [CrossRef]
22. Tao, Y.; Chen, D.; Wei, H.; Yuanhu, P.; Liu, Z.; Huang, L.; Wang, Y.; Xie, S.; Yuan, Z. Development of an accelerated solvent extraction, ultrasonic derivatisation LC-MS/MS method for the determination of the marker residues of nitrofurans in freshwater fish. *Food Addit. Contam. Part A* **2012**, *29*, 736–745. [CrossRef]
23. Wang, K.; Kou, Y.; Wang, M.; Ma, X.; Wang, J. Determination of Nitrofuran Metabolites in Fish by Ultraperformance Liquid Chromatography-Photodiode Array Detection with Thermostatic Ultrasound-Assisted Derivatization. *ACS Omega* **2020**, *5*, 18887–18893. [CrossRef] [PubMed]
24. Palaniyappan, V.; Nagalingam, A.K.; Ranganathan, H.P.; Kandhikuppam, K.B.; Kothandam, H.P.; Vasu, S. Microwave-assisted derivatisation and LC-MS/MS determination of nitrofuran metabolites in farm-raised prawns (*Penaeus monodon*). *Food Addit. Contam. Part A* **2013**, *30*, 1739–1744. [CrossRef]
25. Luo, X.; Sun, Z.; Wang, X.; Yu, Y.; Ji, Z.; Zhang, S.; Li, G.; You, J. Determination of nitrofuran metabolites in marine products by high performance liquid chromatography–fluorescence detection with microwave-assisted derivatization. *New J. Chem.* **2019**, *43*, 2649–2657. [CrossRef]
26. Wang, Q.; Wang, X.F.; Jiang, Y.Y.; Li, Z.G.; Cai, N.; Guan, W.Q.; Huang, K.; Zhao, D.H. Determination of 5-nitro-2-furaldehyde as marker residue for nitrofurazone treatment in farmed shrimps and with addressing the use of a novel internal standard. *Sci. Rep.* **2019**, *9*, 19243. [CrossRef] [PubMed]
27. Zhang, S.; Li, P.; Yan, Z.; Long, J.; Zhang, X. Identification and quantification of nitrofurazone metabolites by ultraperformance liquid chromatography-quadrupole time-of-flight high-resolution mass spectrometry with precolumn derivatization. *Anal. Bioanal. Chem.* **2017**, *409*, 2255–2260. [CrossRef] [PubMed]
28. Zhang, S.; Guo, Y.; Yan, Z.; Sun, X.; Zhang, X. A selective biomarker for confirming nitrofurazone residues in crab and shrimp using ultra-performance liquid chromatography–tandem mass spectrometry. *Anal. Bioanal. Chem.* **2015**, *407*, 8971–8977. [CrossRef] [PubMed]
29. Kaufmann, A.; Butcher, P.; Maden, K. Determination of aminoglycoside residues by liquid chromatography and tandem mass spectrometry in a variety of matrices. *Anal. Chim. Acta* **2012**, *711*, 46–53. [CrossRef]
30. Li, J.; Song, X.; Zhang, M.; Li, E.; He, L. Simultaneous Determination of Aminoglycoside Residues in Food Animal Muscles by Mixed-Mode Liquid Chromatography-Tandem Mass Spectrometry. *Food Anal. Methods* **2018**, *11*, 1690–1700. [CrossRef]
31. Lombardo-Agüí, M.; García-Campaña, A.M.; Cruces-Blanco, C.; Gámiz-Gracia, L. Determination of quinolones in fish by ultra-high performance liquid chromatography with fluorescence detection using QuEChERS as sample treatment. *Food Control* **2015**, *50*, 864–868. [CrossRef]
32. Guidi, L.R.; Tette, P.A.S.; Gloria, M.B.A.; Fernandes, C. A simple and rapid LC-MS/MS method for the determination of amphenicols in Nile tilapia. *Food Chem.* **2018**, *262*, 235–241. [CrossRef]
33. Bortolotte, A.R.; Daniel, D.; de Campos Braga, P.A.; Reyes, F.G.R. A simple and high-throughput method for multiresidue and multiclass quantitation of antimicrobials in pangasius (*Pangasionodon hypophthalmus*) fillet by liquid chromatography coupled with tandem mass spectrometry. *J. Chromatogr. B* **2019**, *1124*, 17–25. [CrossRef] [PubMed]
34. Chen, J.; Wei, Z.; Cao, X.-Y. QuEChERS Pretreatment Combined with Ultra-performance Liquid Chromatography–Tandem Mass Spectrometry for the Determination of Four Veterinary Drug Residues in Marine Products. *Food Anal. Methods* **2019**, *12*, 1055–1066. [CrossRef]
35. Wu, J.; Qian, Y.; Zhang, C.; Zheng, T.; Chen, L.; Lu, Y.; Wang, H. Application of Graphene-based Solid-Phase Extraction Coupled with Ultra High-performance Liquid Chromatography-Tandem Mass Spectrometry for Determination of Macrolides in Fish Tissues. *Food Anal. Methods* **2013**, *6*, 1448–1457. [CrossRef]
36. Sismotto, M.; Paschoal, J.A.R.; Teles, J.A.; de Rezende, R.A.E.; Reyes, F.G.R. A simple liquid chromatography coupled to quadrupole time of flight mass spectrometry method for macrolide determination in tilapia fillets. *J. Food Compos. Anal.* **2014**, *34*, 153–162. [CrossRef]
37. Grande-Martínez, Á.; Moreno-González, D.; Arrebola-Liébanas, F.J.; Garrido-Frenich, A.; García-Campaña, A.M. Optimization of a modified QuEChERS method for the determination of tetracyclines in fish muscle by UHPLC–MS/MS. *J. Pharm. Biomed. Anal.* **2018**, *155*, 27–32. [CrossRef]
38. Shin, D.; Kang, H.-S.; Jeong, J.; Kim, J.; Choe, W.J.; Lee, K.S.; Rhee, G.-S. Multi-residue Determination of Veterinary Drugs in Fishery Products Using Liquid Chromatography-Tandem Mass Spectrometry. *Food Anal. Methods* **2018**, *11*, 1815–1831. [CrossRef]
39. Lopes, R.P.; Reyes, R.C.; Romero-González, R.; Vidal, J.L.M.; Frenich, A.G. Multiresidue determination of veterinary drugs in aquaculture fish samples by ultra high performance liquid chromatography coupled to tandem mass spectrometry. *J. Chromatogr. B* **2012**, *895-896*, 39–47. [CrossRef]

40. Serra-Compte, A.; Alvarez-Munoz, D.; Rodriguez-Mozaz, S.; Barcelo, D. Fast methodology for the determination of a broad set of antibiotics and some of their metabolites in seafood. *Food Chem. Toxicol.* **2017**, *104*, 3–13. [CrossRef]
41. Yuan, G.; Zhu, Z.; Yang, P.; Lu, S.; Liu, H.; Liu, W.; Liu, G. Simultaneous determination of eight nitrofuran residues in shellfish and fish using ultra-high performance liquid chromatography–tandem mass spectrometry. *J. Food Compos. Anal.* **2020**, *92*, 103540. [CrossRef]
42. Jia, W.; Chu, X.; Chang, J.; Wang, P.G.; Chen, Y.; Zhang, F. High-throughput untargeted screening of veterinary drug residues and metabolites in tilapia using high resolution orbitrap mass spectrometry. *Anal. Chim. Acta* **2017**, *957*, 29–39. [CrossRef]
43. Dickson, L.C. Performance characterization of a quantitative liquid chromatography-tandem mass spectrometric method for 12 macrolide and lincosamide antibiotics in salmon, shrimp and tilapia. *J. Chromatogr. B* **2014**, *967*, 203–210. [CrossRef] [PubMed]
44. Habibi, B.; Ghorbel-Abid, I.; Lahsini, R.; Ben Hassen, D.C.; Trabelsi-Ayadi, M. Development and validation of a rapid HPLC method for multiresidue determination of erythromycin, clarithromycin, and azithromycin in aquaculture fish muscles. *Acta Chromatogr.* **2019**, *31*, 109–112. [CrossRef]
45. Orlando, E.A.; Costa Roque, A.G.; Losekann, M.E.; Colnaghi Simionato, A.V. UPLC–MS/MS determination of florfenicol and florfenicol amine antimicrobial residues in tilapia muscle. *J. Chromatogr. B* **2016**, *1035*, 8–15. [CrossRef] [PubMed]
46. Evaggelopoulou, E.N.; Samanidou, V.F. Development and validation of an HPLC method for the determination of six penicillin and three amphenicol antibiotics in gilthead seabream (*Sparus aurata*) tissue according to the European Union Decision 2002/657/EC. *Food Chem.* **2013**, *136*, 1322–1329. [CrossRef] [PubMed]
47. Liu, Y.Y.; Hu, X.L.; Bao, Y.F.; Yin, D.Q. Simultaneous determination of 29 pharmaceuticals in fish muscle and plasma by ultrasonic extraction followed by SPE-UHPLC-MS/MS. *J. Sep. Sci.* **2018**, *41*, 2139–2150. [CrossRef] [PubMed]
48. Li, T.; Wang, C.; Xu, Z.; Chakraborty, A. A coupled method of on-line solid phase extraction with the UHPLCMS/MS for detection of sulfonamides antibiotics residues in aquaculture. *Chemosphere* **2020**, *254*, 126765. [CrossRef] [PubMed]
49. Tang, Y.Y.; Lu, H.F.; Lin, H.Y.; Shin, Y.C.; Hwang, D.F. Development of a Quantitative Multi-Class Method for 18 Antibiotics in Chicken, Pig, and Fish Muscle using UPLC-MS/MS. *Food Anal. Methods* **2012**, *5*, 1459–1468. [CrossRef]
50. Sun, X.; Yang, Y.; Tian, Q.; Shang, D.; Xing, J.; Zhai, Y. Determination of gentamicin C components in fish tissues through SPE-Hypercarb-HPLC-MS/MS. *J. Chromatogr. B* **2018**, *1093–1094*, 167–173. [CrossRef]
51. Gbylik, M.; Posyniak, A.; Mitrowska, K.; Bladek, T.; Zmudzki, J. Multi-residue determination of antibiotics in fish by liquid chromatography-tandem mass spectrometry. *Food Addit. Contam. Part A* **2013**, *30*, 940–948. [CrossRef]
52. Pan, X.D.; Wu, P.G.; Jiang, W.; Ma, B.J. Determination of chloramphenicol, thiamphenicol, and florfenicol in fish muscle by matrix solid-phase dispersion extraction (MSPD) and ultra-high pressure liquid chromatography tandem mass spectrometry. *Food Control* **2015**, *52*, 34–38. [CrossRef]
53. Shen, Q.; Jin, R.; Xue, J.; Lu, Y.; Dai, Z. Analysis of trace levels of sulfonamides in fish tissue using micro-scale pipette tip-matrix solid-phase dispersion and fast liquid chromatography tandem mass spectrometry. *Food Chem.* **2016**, *194*, 508–515. [CrossRef]
54. Wang, Y.; Chen, L. Analysis of malachite green in aquatic products by carbon nanotube-based molecularly imprinted—matrix solid phase dispersion. *J. Chromatogr. B* **2015**, *1002*, 98–106. [CrossRef] [PubMed]
55. Mondal, S.; Xu, J.Q.; Chen, G.S.; Huang, S.M.; Huang, C.Y.; Yin, L.; Ouyang, G.F. Solid-phase microextraction of antibiotics from fish muscle by using MIL-101(Cr)NH$_2$-polyacrylonitrile fiber and their identification by liquid chromatography-tandem mass spectrometry. *Anal. Chim. Acta* **2019**, *1047*, 62–70. [CrossRef] [PubMed]
56. Kung, T.A.; Tsai, C.W.; Ku, B.C.; Wang, W.H. A generic and rapid strategy for determining trace multiresidues of sulfonamides in aquatic products by using an improved QuEChERS method and liquid chromatography-electrospray quadrupole tandem mass spectrometry. *Food Chem.* **2015**, *175*, 189–196. [CrossRef]
57. Munaretto, J.S.; May, M.M.; Saibt, N.; Zanella, R. Liquid chromatography with high resolution mass spectrometry for identification of organic contaminants in fish fillet: Screening and quantification assessment using two scan modes for data acquisition. *J. Chromatogr. A* **2016**, *1456*, 205–216. [CrossRef]
58. Liu, Y.; Yang, H.; Yang, S.; Hu, Q.; Cheng, H.; Liu, H.; Qiu, Y. High-performance liquid chromatography using pressurized liquid extraction for the determination of seven tetracyclines in egg, fish and shrimp. *J. Chromatogr. B* **2013**, *917–918*, 11–17. [CrossRef]
59. Hoff, R.B.; Pizzolato, T.M.; Peralba, M.; Diaz-Cruz, M.S.; Barcelo, D. Determination of sulfonamide antibiotics and metabolites in liver, muscle and kidney samples by pressurized liquid extraction or ultrasound-assisted extraction followed by liquid chromatography-quadrupole linear ion trap-tandem mass spectrometry (HPLC-QqLIT-MS/MS). *Talanta* **2015**, *134*, 768–778. [CrossRef]
60. Kazakova, J.; Fernandez-Torres, R.; Ramos-Payan, M.; Bello-Lopez, M.A. Multiresidue determination of 21 pharmaceuticals in crayfish (*Procambarus clarkii*) using enzymatic microwave-assisted liquid extraction and ultrahigh-performance liquid chromatography-triple quadrupole mass spectrometry analysis. *J. Pharm. Biomed. Anal.* **2018**, *160*, 144–151. [CrossRef] [PubMed]
61. Jansomboon, W.; Boontanon, S.K.; Boontanon, N.; Polprasert, C.; Thi Da, C. Monitoring and determination of sulfonamide antibiotics (sulfamethoxydiazine, sulfamethazine, sulfamethoxazole and sulfadiazine) in imported Pangasius catfish products in Thailand using liquid chromatography coupled with tandem mass spectrometry. *Food Chem.* **2016**, *212*, 635–640. [CrossRef] [PubMed]
62. Chumanee, S.; Sutthivaiyakit, S.; Sutthivaiyakit, P. New Reagent for Trace Determination of Protein-Bound Metabolites of Nitrofurans in Shrimp Using Liquid Chromatography with Diode Array Detector. *J. Agric. Food Chem.* **2009**, *57*, 1752–1759. [CrossRef] [PubMed]

63. Dasenaki, M.E.; Thomaidis, N.S. Multi-residue determination of 115 veterinary drugs and pharmaceutical residues in milk powder, butter, fish tissue and eggs using liquid chromatography-tandem mass spectrometry. *Anal. Chim. Acta* **2015**, *880*, 103–121. [CrossRef]
64. Gaspar, A.F.; Santos, L.; Rosa, J.; Leston, S.; Barbosa, J.; Vila Pouca, A.S.; Freitas, A.; Ramos, F. Development and validation of a multi-residue and multi-class screening method of 44 antibiotics in salmon (*Salmo salar*) using ultra-high-performance liquid chromatography/time-of-flight mass spectrometry: Application to farmed salmon. *J. Chromatogr. B* **2019**, *1118–1119*, 78–84. [CrossRef] [PubMed]
65. Grabicova, K.; Vojs Stanova, A.; Koba Ucun, O.; Borik, A.; Randak, T.; Grabic, R. Development of a robust extraction procedure for the HPLC-ESI-HRPS determination of multi-residual pharmaceuticals in biota samples. *Anal. Chim. Acta* **2018**, *1022*, 53–60. [CrossRef]
66. Guidi, L.R.; Santos, F.A.; Ribeiro, A.C.; Fernandes, C.; Silva, L.H.; Gloria, M.B. A simple, fast and sensitive screening LC-ESI-MS/MS method for antibiotics in fish. *Talanta* **2017**, *163*, 85–93. [CrossRef] [PubMed]
67. Zhao, F.; Gao, X.; Tang, Z.; Luo, X.; Wu, M.; Xu, J.; Fu, X. Development of a simple multi-residue determination method of 80 veterinary drugs in *Oplegnathus punctatus* by liquid chromatography coupled to quadrupole Orbitrap mass spectrometry. *J. Chromatogr. B* **2017**, *1065–1066*, 20–28. [CrossRef]
68. Miossec, C.; Mille, T.; Lanceleur, L.; Monperrus, M. Simultaneous determination of 42 pharmaceuticals in seafood samples by solvent extraction coupled to liquid chromatography-tandem mass spectrometry. *Food Chem.* **2020**, *322*, 126765. [CrossRef] [PubMed]
69. Liang, Y.; Zhou, T. Recent advances of online coupling of sample preparation techniques with ultra high performance liquid chromatography and supercritical fluid chromatography. *J. Sep. Sci.* **2019**, *42*, 226–242. [CrossRef]
70. Ma, J.; Fan, S.; Sun, L.; He, L.; Zhang, Y.; Li, Q. Rapid analysis of fifteen sulfonamide residues in pork and fish samples by automated on-line solid phase extraction coupled to liquid chromatography–tandem mass spectrometry. *Food Sci. Hum. Wellness* **2020**, *9*, 363–369. [CrossRef]
71. Hurtado de Mendoza, J.; Maggi, L.; Bonetto, L.; Rodríguez Carmena, B.; Lezana, A.; Mocholí, F.A.; Carmona, M. Validation of antibiotics in catfish by on-line solid phase extraction coupled to liquid chromatography tandem mass spectrometry. *Food Chem.* **2012**, *134*, 1149–1155. [CrossRef]
72. Bonfiglio, R.; King, R.C.; Olah, T.V.; Merkle, K. The Effects of Sample Preparation Methods on the Variability of the Electrospray Ionization Response for Model Drug Compounds. *Rapid Commun. Mass Spectrom.* **1999**, *13*, 1175–1185. [CrossRef]
73. Van Eeckhaut, A.; Lanckmans, K.; Sarre, S.; Smolders, I.; Michotte, Y. Validation of bioanalytical LC-MS/MS assays: Evaluation of matrix effects. *J. Chromatogr. B* **2009**, *877*, 2198–2207. [CrossRef] [PubMed]
74. Kim, D.; Kim, B.; Hyung, S.-W.; Lee, C.H.; Kim, J. An optimized method for the accurate determination of nitrofurans in chicken meat using isotope dilution–liquid chromatography/mass spectrometry. *J. Food Compos. Anal.* **2015**, *40*, 24–31. [CrossRef]
75. Ferrer, C.; Lozano, A.; Aguera, A.; Giron, A.J.; Fernandez-Alba, A.R. Overcoming matrix effects using the dilution approach in multiresidue methods for fruits and vegetables. *J. Chromatogr. A* **2011**, *1218*, 7634–7639. [CrossRef] [PubMed]
76. Lee, S.; Kim, B.; Kim, J. Development of isotope dilution-liquid chromatography tandem mass spectrometry for the accurate determination of fluoroquinolones in animal meat products: Optimization of chromatographic separation for eliminating matrix effects on isotope ratio measurements. *J. Chromatogr. A* **2013**, *1277*, 35–41. [CrossRef] [PubMed]
77. Douny, C.; Widart, J.; De Pauw, E.; Silvestre, F.; Kestemont, P.; Tu, H.T.; Phuong, N.T.; Maghuin-Rogister, G.; Scippo, M.-L. Development of an analytical method to detect metabolites of nitrofurans. *Aquaculture* **2013**, *376*, 54–58. [CrossRef]

Article

Green and Simple Extraction of Arsenic Species from Rice Flour Using a Novel Ultrasound-Assisted Enzymatic Hydrolysis Method

Xiao Li [1], Qian Ma [1], Chao Wei [1], Wei Cai [2], Huanhuan Chen [1], Rui Xing [3] and Panshu Song [1,*]

[1] Division of Chemical Metrology and Analytical Science, National Institute of Metrology, Beijing 100029, China; lixiao@nim.ac.cn (X.L.); maqian@nim.ac.cn (Q.M.); weichao@nim.ac.cn (C.W.); chenhuanhuan@nim.ac.cn (H.C.)

[2] Department of Chemical Analysis and Biomedicine, Beijing Institute of Metrology, Beijing 100029, China; caiw@bjjl.cn

[3] College of Life Science and Technology, Beijing University of Chemical Technology, Beijing 100029, China; xingruii2022@163.com

* Correspondence: songpsh@nim.ac.cn; Tel.: +86-10-6452-4781

Abstract: It is well established that arsenic (As) has many toxic compounds, and in particular, inorganic As (iAs) has been classified as a type-1 carcinogen. The measuring of As species in rice flour is of great importance since rice is a staple of the diet in many countries and a major contributor to As intake in the Asian diet. In this study, several solvents and techniques for the extraction of As species from rice flour samples prior to their analysis by HPLC-ICP-MS were investigated. The extraction methods were examined for their efficiency in extracting various arsenicals from a rice flour certified reference material, NMIJ-7532a, produced by the National Metrology Institute of Japan. Results show that ultrasound-assisted extraction at 60 °C for 1 h and then heating at 100 °C for 2.5 h in the oven using a thermostable α-amylase aqueous solution was highly effective in liberating the arsenic species. The recoveries of iAs and dimethylarsinic acid (DMA) in NMIJ-7532a were 99.7% ± 1.6% ($n = 3$) and 98.1% ± 2.3% ($n = 3$), respectively, in comparison with the certificated values. Thus, the proposed extraction method is a green procedure that does not use any acidic, basic, or organic solvents. Moreover, this extraction method could effectively maintain the integrity of the native arsenic species of As(III), As(V), monomethylarsonate (MMA), DMA, arsenobetaine (AsB) and arsenocholine (AsC). Under the optimum extraction, chromatography and ICP-MS conditions, the limits of detection (LOD) obtained were 0.47 ng g^{-1}, 1.67 ng g^{-1} and 0.80 ng g^{-1} for As(III), As(V) and DMA, respectively, while the limits of quantification (LOQ) achieved were 1.51 ng g^{-1}, 5.34 ng g^{-1} and 2.57 ng g^{-1} for As(III), As(V) and DMA, respectively. Subsequently, the proposed method was successfully applied to As speciation analysis for several rice flour samples collected from contaminated areas in China.

Keywords: arsenic speciation; rice flour; enzymatic hydrolysis; ultrasound-assisted extraction; HPLC-ICP-MS

1. Introduction

Arsenic (As) is toxic and has been classified as a human carcinogen by the International Agency for Research on Cancer (IRAC) [1]. There are several naturally occurring chemical forms of arsenic, such as inorganic compounds of As(III) and As(V) and organic species of methylarsonic acid (MMA), dimethylarsinic acid (DMA), arsenobetaine (AsB), arsenocholine (AsC), etc. Of these species, inorganic arsenic (iAs) is classified as a type-1 carcinogen, with lethal dose 50 (LD$_{50}$) values in the range of 15–42 mg kg^{-1} for As(III) and 20–200 mg kg^{-1} for As(V); following that, the organic forms of arsenic MMA and DMA are possibly carcinogenic to humans (type 2B), with LD$_{50}$ values in the range of

700–1800 mg kg^{-1} and 1200–2600 mg kg^{-1}, respectively [2,3]. Therefore, the Food and Agriculture Organization/World Health Organization (FAO/WHO) have recommended a provisional tolerable weekly intake (PTWI) of iAs at 15 µg kg^{-1} of body weight [4].

Arsenic is gradually accumulated in crops from the water, soil and air owing to human industrial activities [5,6]. As a staple of the diet in Asia, rice contributes more arsenic to the Asian diet than all other agricultural products [7]. Moreover, it has been reported that the arsenic concentration in rice is 10 times higher in comparison with that in other crops under flooded soil conditions [8,9]. As the largest rice-producing and -consuming country in the world, China has regulated the iAs concentration in rice to be not more than 0.2 mg kg^{-1} through its National Medical Products Administration [10,11]. Thus, it is necessary to investigate the concentrations of arsenic species in rice to ensure that rice can be safely consumed.

The studies of arsenic speciation analysis have grown rapidly in recent years and are generally conducted using high-performance liquid chromatography coupled to inductively coupled plasma mass spectrometry (HPLC-ICPMS) [12–16]. Prior to that, speciation analysis usually requires the extraction of the species from complex matrices into solvents before they can be identified and measured. This is also one of the key steps for accurate determination of As species in rice flour samples, as incomplete extraction of species would lead to erroneous results [17]. In previous studies, a number of procedures utilizing different extraction solvents, such as water, diluted nitric acid, methanol–water or nitric acid–hydrogen peroxide mixtures, with the extraction techniques of mechanical agitation, sonication, microwave heating or pressurized liquid, have been implemented for extraction of As species from rice samples [18–23]. Narukawa et al. have reported the extraction of As from rice using water at 90–100 °C. However, the extraction efficiency of As(III) varied depending on the rice sample analyzed; thus, the total arsenic extracted was in the range from 80% to 100% of the arsenic present. Therefore, in order to achieve 100% extraction of As species, it is still necessary to optimize the extraction conditions for rice samples. Moreover, the use of acidic or organic solvents would increase occupational risk and the amount of hazardous waste, which are the main disadvantages of other reported extraction methods [24].

In the present study, we developed and optimized a method based on ultrasound-assisted enzymatic hydrolysis, which provided for the quantitative, green and simple extraction of arsenicals in rice flour samples, and hence facilitated As speciation analysis using HPLC-ICP-MS. Several extraction media and techniques were evaluated based on the total arsenic extracted and the extraction efficiencies for different chemical forms of native As species in a rice flour reference material NMIJ-7532a. The optimum method is based on hydrolyzing the starch and other components in rice by mainly using α-amylase. Ultrasonic and heating techniques were implemented to enhance the extraction efficiency and shorten the time consumption. Once optimized, the method was further validated by analyzing two other rice flour reference materials, NMIJ-7501a and NIST SRM 1568b. Subsequently, arsenic speciation in a variety of real rice flour samples was carried out.

2. Experimental Section

2.1. Reagents and Standards

High-purity nitric acid was purchased from Sigma-Aldrich Inc. (St. Louis, MO, USA) and used for sample digestion. Ultra-pure water with a resistivity of 18.2 MΩ·cm was produced by a Milli-Q system (Millipore Corp., Burlington, MA, USA) and used for the preparation of all samples and standard solutions. The thermostable α-amylase from Sigma-Aldrich Inc. (St. Louis, MO, USA) was used. For the preparation of the mobile phase, dibasic ammonium hydrogen phosphate (purity ≥ 99.0%) was supplied by Sigma-Aldrich Inc. (St. Louis, MO, USA). HPLC-grade methanol was purchased from Thermo Fisher Scientific (Waltham, MA, USA). The mobile phase was degassed and filtered before use. The syringe filter (0.22 µm) PTFE PVDF was from Jinteng Experimental Equipment (Tianjin, China).

The certified reference materials of arsenite (As(III), GBW08666), arsenate (As(V), GBW08667), monomethylarsonic acid (MMA, GBW08668), dimethylarsinic acid (DMA, GBW08669), arsenobetaine (AsB, GBW08670) and arsenocholine bromide (AsC, GBW08671) were all produced by the National Institute of Metrology (Beijing, China). The concentrations of all standard solutions are SI-traceable.

Certified reference materials of As species in rice flour, NMIJ-7532a, from the National Metrology Institute of Japan (NMIJ, Tsukuba, Japan) were selected for method development. The information and certified values of the total As, iAs and DMA were 0.320 ± 0.010 mg/kg, 0.298 ± 0.008 mg/kg and 0.0186 ± 0.0008 mg/kg, respectively.

2.2. Intrumentation

A Mars5 microwave system (CEM, Matthews, NC, USA) was used for the extraction of the total arsenic. Ultrasound-assisted extractions were carried out by an ultrasonic bath (KQ-500GDV, Kun Shan Ultrasonic Instruments Co., Ltd, Kunshan, Jiangsu, China). Shaking was performed on a WS20 shaking incubator (Wiggens, Straubenhardt, Germany). A DKN612C forced convection oven (Yamato Scientific Co., Ltd., Tokyo, Japan) was utilized as the heating device during digestion. A universal 320R centrifuge (Hettich, Tuttlingen, Germany) was employed for the centrifugation of the extracts obtained from the samples.

An ICP-MS (8800, Agilent Technologies, Santa Clara, CA, USA) equipped with a collision cell, a Scott double pass spray chamber and a PFA nebulizer was used for the determination of As. To reduce some polyatomic molecular interferences ($^{40}Ar^{35}Cl^+$, $^{59}Co^{+16}O^+$) in $^{75}As^+$ analysis, the collision cell gas of O_2 was set as 25%. In this case, the mass monitored was $m/z = 91$ for AsO^+. The spray chamber temperature, collision gas flow rate and $^{91}AsO^+$ signal integration time were optimized to obtain the best limit of detection (LOD) of As. Typical operating parameters for the ICP-MS are summarized in Table 1.

Table 1. Operating conditions of the HPLC-ICP-MS system.

ICP-MS
RF powder: 1500 W
Carrier gas: 1.0 mL/min
Reaction gas: O_2, 25%
Isotope monitored: $^{91}AsO^+$
Integration time: 0.1 s (spectrum) per point
Points per peak: 3
HPLC
Column: PRP-X100 anion exchange
Dimensions: 250 mm × 4.1 mm, particle size: 10 μm
Mobile phase: 20 mM $(NH_4)_2HPO_4$, pH 6.0
Injection volume: 20 μL
Flow rate: 1.2 mL/min
Mode: Isocratic

For arsenic speciation, chromatographic separations were performed by an Agilent 1290 HPLC system (Santa Clara, CA, USA). A Hamilton PRP-X100 anion exchange column (Grace, Belgium) was used as the stationary phase. Guard columns of the same stationary phase were connected in front of the separation columns. All the columns were preconditioned according to the manufacturer's instructions before use. The pH of the mobile phase used in LC was adjusted by a Mettler Toledo FiveEasy Plus pH meter (Zurich, Switzerland).

2.3. Samples

Rice samples were obtained by pooling individual samples ($n = 3$) of different varieties grown in contaminated areas in China. The rice flour certified reference materials were NMIJ-7502a, or brown rice obtained by NMIJ (NMIJ, Tsukuba, Japan), and NIST SRM

1568b, obtained by the National Institute of Standards and Technology (NIST, Gaithersburg, MD, USA).

2.4. Total Arsenic Determination

The total arsenic in the reference material NMIJ-7532a and the test samples was determined in accordance with the following procedures. Three replicate portions of each sample (0.5 g dry weight) were weighed into Teflon microwave vessels. High-purity nitric acid (5 mL) was added into the vessels, and the mixtures were predigested for 6 h or overnight in a clean air hood at an ambient temperature (20 °C). This step could effectively prevent the samples with high starch from bumping during digestion. After that, the samples were placed in the microwave digestion system. The microwave program was as follows: 15 min ramp to 150 °C, held for 5 min at 150 °C; then, 10 min ramp to 175 °C, held for 5 min at 175 °C; finally, 10 min ramp to 195 °C, held for 35 min at 195 °C. The digested samples were cooled to room temperature and then transferred to 50-mL polypropylene test tubes, which were gravimetrically diluted with ultra-pure water for ICP-MS analysis. For sample digestion, each digestion batch consisted of two procedural blanks and triplicates of the reference material NMIJ-7532a.

2.5. Evaluation of Extraction Methods

Table 2 shows the extraction methods evaluated in this study. The list includes acidic and enzymatic extractions. Different extraction temperatures, times and devices were also evaluated to provide an optimum extraction procedure for the analysis of As species in rice flour. The reference material of rice flour, NMIJ-7532a, was used to examine their efficiencies in extracting iAs and DMA. The sample weight was 0.5 g for all procedures. Each extraction procedure was repeated at least three times. Procedural blanks were included in each extraction batch. Samples were kept in the refrigerator after filtration prior to HPLC-ICP-MS measurements.

Table 2. Extraction conditions evaluated in this study.

Extraction	Procedure	Solvent/Solution	Extraction Temperature and Time	Heating Device
Acidic	A-1	10 mL 1% (v/v) HNO_3	stand overnight; 80 °C, 2.5 h	Oven
	A-2	10 mL 1% (v/v) HNO_3	stand overnight; 90 °C, 2.5 h	Oven
	A-3	10 mL 1% (v/v) HNO_3	stand overnight; 100 °C, 2.5 h	Oven
	A-4	10 mL 1% (v/v) HNO_3	stand overnight; 100 °C, 0.5 h	Oven
	A-5	10 mL 1% (v/v) HNO_3	stand overnight; 100 °C, 1.5 h	Oven
	A-6	10 mL 1% (v/v) HNO_3	stand overnight; 100 °C, 2.5 h	Oven
	A-7	10 mL 1% (v/v) HNO_3	stand overnight; 100 °C, 3.5 h	Oven
Enzymatic	B-1	5 mL 10 mg/mL α-amylase	60 °C, overnight	Shaking incubator
	B-2	5 mL 10 mg/mL α-amylase	60 °C, overnight; 100 °C, 2.5 h	Shaking incubator Oven
Ultrasound-assisted enzymatic	C-1	5 mL 10 mg/mL α-amylase	60 °C, 1 h	Ultrasonic bath
	C-2	5 mL 10 mg/mL α-amylase	60 °C, 1 h; 100 °C, 2.5 h	Ultrasonic bath Oven

Procedure A (Acidic extraction)—A total of 10 mL of 1% (v/v) HNO_3 was added to the samples, which were then predigested overnight in a clean air hood at an ambient temperature. The next day, the samples were placed in the oven for digestion and shaken every 30 min. Different extraction times of 0.5 h to 3.5 h under various temperatures from 80 °C to 100 °C were examined. After each extraction, the samples were centrifuged at 8000 rpm for 10 min under 4 °C. The supernatants were then filtered through a 0.22 μm filter.

Procedure B (Enzymatic extraction)—A total of 5 mL of 10 mg/mL α-amylase aqueous solution were added to the samples, which were then treated in a shaker at 60 °C overnight. The next day, the samples were placed in the oven for 2.5 h under 100 °C for digestion (shaken every 30 min). The extracts were centrifuged and filtered as described in procedure A.

Procedure C (Ultrasound-assisted enzymatic extraction)—A total of 5 mL of 10 mg/mL α-amylase aqueous solution were added to the samples, which were then placed in ultrasound equipment for 1 h at 80 W. Then, further digestion was performed in the oven for 2.5 h at 100 °C (shaken every 30 min). The extracts were treated as described above.

2.6. Determination of Arsenic Species

Arsenic species were determined by HPLC-ICP-MS using a strong anion exchange column fitted with a matching guard column filled with the identical phase. The operating conditions of the HPLC-ICP-MS system are shown in Table 1. The arsenic species were identified by retention time matching with the mixed standard solution consisting of As(III), As(V), MMA, DMA, AsB and AsC substances as external standards.

2.7. Analyte Quantification

For the total arsenic analysis using ICP-MS, the standard additions method was carried out to correct the matrix effect. The calibration standards were prepared by serially diluting the 1 mg kg^{-1} arsenic stock solution in 6% HNO$_3$ to obtain final concentrations of 0.5 to 5 ng g^{-1}. For quantifying As species by HPLC-ICP-MS, the external calibration method was used with a standard solution containing 10 ng g^{-1} of each arsenicals in 1% HNO$_3$. All calibration standards were prepared gravimetrically and daily before use. Data acquisition, chromatographic peak integration and quantification were all carried out using the Agilent MassHunter software. Microsoft Excel was used for further calculations.

3. Results and Discussion

3.1. Determination of Total Arsenic by Microwave Digestion

The determinations of the total arsenic in the certified reference materials and rice flour samples were performed using ICP-MS after microwave digestion. The results are reported in Table 3. The total As ($n = 3$) determined in reference materials NMIJ-7532a, NMIJ 7502a and NIST SRM 1568b were 0.316 ± 0.012 mg kg^{-1}, 0.112 ± 0.003 mg kg^{-1} and 0.288 ± 0.006 mg kg^{-1}, respectively, which were in good agreement with their certified values. The total As concentrations of the three rice flour samples ($n = 3$) were 0.157 ± 0.003 mg kg^{-1}, 0.482 ± 0.008 mg kg^{-1} and 0.378 ± 0.008 mg kg^{-1}, respectively. These values were further used for calculating the recoveries of the total As during sample analysis under the optimum extraction procedure, since no reference value was available for them.

Table 3. Total arsenic in the certified reference materials and rice flour samples determined by ICP-MS after MW.

Code	Description	Total As (mg kg^{-1})	Certified Value (mg kg^{-1})
NMIJ 7532a	Rice flour from NMIJ	0.316 ± 0.012 [a]	0.320 ± 0.010
NMIJ 7502a	Rice flour from NMIJ	0.112 ± 0.003	0.109 ± 0.005
NIST 1568b	Rice flour from NIST	0.288 ± 0.006	0.285 ± 0.014
S1	Chinese brown rice flour	0.157 ± 0.003	-
S2	Chinese brown rice flour	0.482 ± 0.008	-
S3	Chinese brown rice flour	0.378 ± 0.008	-

[a] The standard deviation has been obtained for $n = 3$.

3.2. Determination of Arsenic Species

The As speciation was analyzed by the HPLC-ICP-MS method. The chromatograms obtained with the mixed standard solutions (100 ng g^{-1} as As) are displayed in Figure 1. As can be seen, the six arsenic species of As(III), As(V), MMA, DMA, AsB and AsC were well separated on the PRP-X100 column using an isocratic gradient (conditions see Table 1). Quantification was based on peak area measurements.

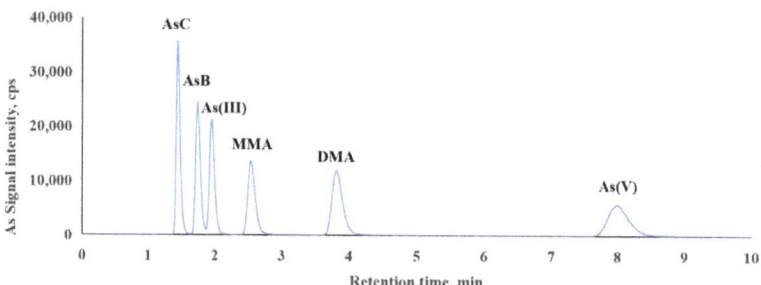

Figure 1. HPLC-ICP-MS chromatogram of a mixture of standards at a concentration of 100 ng g^{-1} as As.

In the measurement of As by ICP-MS, it is well known that ^{40}Ar^{35}Cl$^+$ and ^{59}Co^{16}O$^+$ with m/z = 75 causes interferences for ^{75}As$^+$ analysis. Therefore, oxygen was used as reaction gas and resulted in detection of arsenic as AsO$^+$ at m/z = 91 in this study. The reproducibility of peak areas for the mixture of standards at a concentration of 10 ng g^{-1} as As was measured over a period of approximately 2 h (n = 10). The relative standard deviations were 2.28%, 1.56%, 2.14% and 2.29% for As(III), As(V), MMA and DMA, respectively.

3.3. Optimization of Extraction Methods

The rice flour certified reference material NMIJ-7532a was used to compare the extraction efficiencies of the two solvents and three extraction techniques. The speciation analysis of arsenic was carried out by HPLC-ICP-MS. The concentrations of the extracted arsenic species and total As were directly compared with the values stated in their certificates. The obtained results are reported in Table 4.

Table 4. Results of arsenic speciation of NMIJ 7532a.

Procedure	iAs (mg kg^{-1})	DMA (mg kg^{-1})	Total (mg kg^{-1})
A-1	0.253 ± 0.005 (84.8%) [a]	0.0163 ± 0.0003 (87.9%)	0.269 ± 0.005 (84.1%)
A-2	0.284 ± 0.006 (95.3%)	0.0171 ± 0.0003 (92.1%)	0.301 ± 0.006 (94.1%)
A-3	0.299 ± 0.006 (100.3%)	0.0182 ± 0.0004 (97.8%)	0.317 ± 0.006 (99.1%)
A-4	0.264 ± 0.006 (88.6%)	0.0170 ± 0.0003 (91.4%)	0.281 ± 0.006 (87.8%)
A-5	0.288 ± 0.006 (96.6%)	0.0190 ± 0.0004 (102.2%)	0.307 ± 0.006 (95.9%)
A-6	0.300 ± 0.006 (100.7%)	0.0190 ± 0.0004 (102.2%)	0.319 ± 0.006 (99.7%)
A-7	0.298 ± 0.006 (100.0%)	0.0180 ± 0.0004 (96.8%)	0.316 ± 0.006 (98.8%)
B-1	0.220 ± 0.005 (73.7%)	0.0133 ± 0.0003 (71.4%)	0.233 ± 0.005 (72.8%)
B-2	0.240 ± 0.005 (80.4%)	0.0145 ± 0.0003 (78.2%)	0.254 ± 0.005 (79.4%)
C-1	0.278 ± 0.006 (93.2%)	0.0163 ± 0.0003 (87.5%)	0.294 ± 0.006 (91.9%)
C-2	0.297 ± 0.006 (99.7%)	0.0182 ± 0.0004 (98.1%)	0.315 ± 0.006 (98.5%)
Certified value	0.298 ± 0.008	0.0186 ± 0.0008	0.320 ± 0.010

[a] Recovery displayed in parentheses.

3.3.1. Extraction by Acidic Solvent

Diluted solutions of HNO_3 have been widely used to extract arsenic from samples in various matrices. As illustrated in GB/5009.11-2014, published by the National Health Commission of China [25], the optimum extraction conditions for arsenic in rice is using 1% HNO_3 at 90 °C for 2.5 h. However, different extraction temperatures and times were reported in the literature [18,19,21,26]. For verifying the optimum extraction protocol for arsenic species in rice flour, the extraction efficiencies of 1% HNO_3 were evaluated at different heating temperatures using a forced convection oven (80 °C, 90 °C and 100 °C) and extraction times of 0.5 to 3.5 h in this study. As can be seen from Table 4, under the same extraction time of 2.5 h, the extraction yields increased from 84.8% to 100.3% for iAs and from 87.9% to 97.8% for DMA when the heating temperature increased from 80 °C to 100 °C. It was also possible to achieve the complete extraction of iAs and DMA in rice flour in 2.5 h at 100 °C. No significant advantages in extraction efficiencies were observed for either iAs or DMA when the extraction time increased from 2.5 h to 3.5 h under 100 °C. Hence, the extraction of arsenic species in rice flour with 1% HNO_3 under 100 °C is recommended instead of 90 °C for 2.5 h.

3.3.2. Enzymatic Extraction by Shaking

As reported in the literature, α-amylase hydrolyses the α-1,4-linkage of starch, which is the major component in rice of up to 78% [27]. This process could increase the solubility of the proteins by liberating the starch-bound proteins and hence facilitating the arsenic extraction. In order to establish an effective extraction method for As species in rice flour using α-amylase, the extraction efficiency was first evaluated with α-amylase aqueous solution by shaking treatment. As shown in Table 4, the extractions of iAs and DMA were only 73.7% and 71.4%, respectively, after being treated with α-amylase aqueous solution and left in a shaking incubator at 60 °C overnight. Increased extraction efficiencies were obtained of 80.4% and 78.2% for iAs and DMA, respectively, when the extracts were further heated in the oven under 100 °C for 2.5 h, although the method for enzymatic extraction using α-amylase still needs to be optimized for the effective extraction of As species in rice flour.

3.3.3. Ultrasound-Assisted Enzymatic Extraction

In order to enhance the extraction efficiencies of As species and reduce sample treatment time, the ultrasonic technique was employed to accelerate the enzymatic hydrolysis activity. The results are summarized in Table 4. The detected iAs and DMA were increased to 93.2% and 87.5%, respectively, after 1 h of sonication with α-amylase aqueous solution as the extraction media. Extraction efficiencies of approximately 100% could be achieved for both iAs and DMA when the samples were further heated in the oven for 2.5 h under 100 °C after ultrasound-assisted extraction. These results suggested that the reactivity of α-amylase was enhanced with physical processing by acoustic cavitation, and hence, accelerated the reaction rates of enzymatic hydrolysis. Moreover, the heating process is necessary for achieving the complete extraction of As species in rice flour.

The optimum conditions for ultrasound-assisted enzymatic extraction were as follow: 5 mL of 10 mg/mL α-amylase as the extraction solvent, sonication treatment for 1 h at 60 °C; then, heating in the oven for 2.5 h under the extraction temperature of 100 °C. Thus, an effective method for extracting As species from rice flour was developed, which was more green, rapid and simple in comparison with the conventional acidic extraction method.

3.3.4. Stability of Arsenic Species

The maintenance of the native chemical forms of As is an essential requirement for As speciation analysis during the extraction process. Hence, to evaluate the preservation of As species by the ultrasound-assisted enzymatic extraction method, the recoveries of As(III), As(V), MMA, DMA, AsB and AsC were investigated. The rice flour reference material NMIJ-7532a was spiked with each of the As species' standards (10 and 100 ng g^{-1} as As), and their recoveries were determined following the proposed optimum extraction and quantification procedures. As displayed in Table 5, the recoveries of all these species were nearly 100%. These results demonstrated that no chemical alteration happened to these As species during the proposed ultrasound-assisted enzymatic extraction process.

Table 5. Recoveries of As species during ultrasound-assisted enzymatic extraction.

Concentration of Spiked Standards	Recovery (%)					
	As(III)	As(V)	MMA	DMA	AsB	AsC
10 ng g^{-1}	99.6	100.7	99.5	98.5	99.4	98.0
100 ng g^{-1}	100.2	101.3	100.3	98.9	100.1	99.2

3.3.5. Limits of Detection and Quantification

The limits of detection (LOD) and quantification (LOQ) of the optimized protocols were determined based on the standard deviation of replicate (n = 10) analyses of a blank solution of 1% HNO$_3$. The results are listed in Table 6.

Table 6. Limits of detection and quantification of arsenic species (ng g^{-1}).

Species	As(III)	As(V)	MMA	DMA	AsB	AsC
LOD	0.47	1.67	0.71	0.80	0.32	0.21
LOQ	1.51	5.34	2.26	2.57	1.01	0.66

3.4. Speciation Analysis of Arsenicals in Rice Flour Samples

The ultrasound-assisted enzymatic extraction was applied to determine the arsenic species in five rice flour samples in this study, including two rice flour reference materials and three real rice samples (S1 to S3) collected from the contaminated areas in China. The obtained results are summarized in Table 7. For reference materials NMIJ-7502a and NIST SRM 1586a, the detected amounts of iAs, MMA and DMA and the total As recovered show no significant differences from their certificated values. These results further validated the reliability of the proposed enzymatic extraction method. In addition, for all of the three real rice samples, the sums of each As species were in good agreement with the total arsenic concentrations that were determined by ICP-MS after microwave-assisted digestion.

Table 7. Results of Arsenic species in rice flour samples determined using ultrasound-assisted enzymatic extraction.

Code	iAs (mg kg^{-1})	MMA (mg kg^{-1})	DMA (mg kg^{-1})	Sum (mg kg^{-1})
NMIJ 7502a	0.096 ± 0.003 (98.0%) [a]	_ [b]	0.0130 ± 0.0004 (100.7%)	0.109 (100.0%)
NIST 1568b	0.089 ± 0.003 (96.7%)	0.0118 ± 0.004 (101.7%)	0.181 ± 0.006 (100.5%)	0.282 (98.9%)
S1	0.149 ± 0.005	-	0.0061 ± 0.0002	0.155 (98.7%)
S2	0.440 ± 0.015	0.0032 ± 0.0002	0.041 ± 0.001	0.484 (100.4%)
S3	0.365 ± 0.012	-	0.0092 ± 0.0003	0.374 (98.9%)

[a] The recovery displayed in parentheses for each As species was calculated by dividing its certified value; the recovery displayed in parentheses for the sum of As was obtained by dividing the certified value or total digestion of As. [b] Not detected.

As the results show in Table 7, the relative amount of iAs ranged from 90.9% to 97.6% of the sum of the As species in rice samples S1–S3. Most of the remaining arsenical was DMA. In samples S1 and S3, MMA was not detected. These results confirm that rice is a bio-accumulative plant for the more toxic As species. Furthermore, the concentrations of iAs in samples S2 and S3 were 0.440 ± 0.015 mg kg^{-1} and 0.365 ± 0.012 mg kg^{-1}, respectively, which are higher than the nationally set safety standards of China for iAs in rice (0.2 mg kg^{-1}). Thus, the importance of monitoring the contents of As species in rice should be further addressed to ensure that rice can be safely consumed.

4. Conclusions

In this study, various conditions for extracting arsenic species from rice flour samples were investigated. HPLC-ICP-MS was used as the analytical method for the determination of As species. Enzymatic hydrolysis based on the use of α-amylase in aqueous media, in conjunction with ultrasound treatment, has been found to be a green, rapid and simple extraction method for the determination of As species in rice flour. The extraction efficiencies, 96.7~99.7% for iAs and 98.1~100.7% for DMA, were obtained by analyzing three rice flour reference materials (NMIJ-7532a, NMIJ-7502a and NIST SRM 1568a). Moreover, the proposed ultrasound-assisted enzymatic extraction method showed advantages in preserving the chemical integrity of native arsenic species of As(III), As(V), MMA, DMA, AsB and AsC with nearly 100% recoveries.

The novel extraction method could also meet the analytical needs for accurate quantification of As species in real rice samples. Our results indicated that more than 90% of As was found to be present in its inorganic form, which is more toxic, in the tested rice samples from China, whereas the remainder was mainly DMA. In this instance, persistent monitoring of the content of As species in rice products is necessary to prevent the dietary exposure to iAs for populations consuming a predominantly rice-based diet.

Author Contributions: Methodology, project administration, X.L.; validation, Q.M.; data curation, C.W.; data curation, formal analysis, W.C.; data curation, investigation, H.C.; formal analysis, R.X.; writing—review and editing, P.S. All authors have read and agreed to the published version of the manuscript.

Funding: This research was funded by the National Key R&D Program of China, grant number 2019YFC1604803.

Data Availability Statement: Not applicable.

Conflicts of Interest: The authors declare no conflict of interest.

References

1. International Agency for Research on Cancer (IARC). IARC Monographs on the Evaluation of Carcinogenic Risks to Humans; Volume 100C: Arsenic, Metals, Fibers, and Dusts. Available online: http://monographs.iarc.fr/ENG/Monographs/vol100C/mono100C-6.pdf (accessed on 13 December 2016).
2. Yim, S.R.; Park, G.Y.; Lee, K.W.; Chung, M.S.; Shim, S.M. Determination of total arsenic content and arsenic speciation in different types of rice. *Food Sci. Biotechnol.* **2017**, *26*, 293–298. [CrossRef]
3. Akter, K.F.; Owens, G.; Davey, D.E.; Naidu, R. Arsenic Speciation and Toxicity in Biological Systems. *Rev. Environ. Contam. Toxicol.* **2005**, *184*, 97–149. [CrossRef]
4. *WHO Evaluation of Certain Food Additives and Contaminants (1989) 33rd Report of the Joint FAO/WHO Expert Committee on Food Additives*; WHO Technical Report Series 776; WHO: Geneva, Switzerland, 1988.
5. Ma, J.F.; Yamaji, N.; Mitani, N.; Xu, X.Y.; Su, Y.H.; McGrath, S.P.; Zhao, F.J. Transporters of arsenite in rice and their role in arsenic accumulation in rice grain. *Proc. Natl. Acad. Sci. USA* **2008**, *105*, 9931–9935. [CrossRef]
6. Williams, P.N.; Price, A.H.; Raab, A.; Hossain, S.A.; Feldmann, J.; Meharg, A.A. Variation in Arsenic Speciation and Concentration in Paddy Rice Related to Dietary Exposure. *Environ. Sci. Technol.* **2005**, *39*, 5531–5540. [CrossRef]
7. Lombi, E.; Scheckel, K.G.; Pallon, J.; Carey, A.M.; Zhu, Y.G.; Meharg, A.A. Speciation and distribution of arsenic and localization of nutrients in rice grains. *New Phytol.* **2009**, *184*, 193–201. [CrossRef]
8. Zhu, Y.G.; Sun, G.X.; Lei, M.; Teng, M.; Liu, Y.X.; Chen, N.C.; Wang, L.H.; Carey, A.M.; Deacon, C.; Raab, A.; et al. High percentage inorganic arsenic content of mining impact and monimnacted chinese rice. *Environ. Sci. Technol.* **2008**, *42*, 5008–5013. [CrossRef]
9. Williams, P.N.; Raab, A.; Feldmann, J.; Meharg, A.A. Market Basket Survey Shows Elevated Levels of As in South Central, U.S. Processed Rice Compared to California: Consequences for Human Dietary Exposure. *Environ. Sci. Technol.* **2007**, *41*, 2178–2183. [CrossRef]
10. Li, X.; Xie, K.; Yue, B.; Gong, Y.Y.; Shao, Y.; Shang, X.; Wu, Y. Inorganic arsenic contamination of rice from Chinese major rice-producing areas and exposure assessment in Chinese population. *Sci. China Chem.* **2015**, *58*, 1898–1905. [CrossRef]

11. *GB2762–2017*; National Food Safety Standard-Limit of Pollutants in Food. National Health and Family Planning Commission of China: Beijing, China, 2017.
12. Sanz, E.; Muñoz-Olivas, R.; Dietz, C.; Sanz, J.; Cámara, C. Alternative extraction methods for arsenic speciation in hair using ultrasound probe sonication and pressurised liquid extraction. *J. Anal. At. Spectrom.* **2007**, *22*, 131–139. [CrossRef]
13. Kubachka, K.M.; Conklin, S.D.; Smith, C.C.; Castro, C. Quantitative Determination of Arsenic Species from Fruit Juices Using Acidic Extraction with HPLC-ICPMS. *Food Anal. Methods* **2019**, *12*, 2845–2856. [CrossRef]
14. Wolle, M.M.; Conklin, S.D. Speciation analysis of arsenic in seafood and seaweed: Part I—Evaluation and optimization of methods. *Anal. Bioanal. Chem.* **2018**, *410*, 5675–5687. [CrossRef]
15. Maher, W.A.; Ellwood, M.J.; Krikowa, F.; Raber, G.; Foster, S. Measurement of arsenic species in environmental, biological fluids and food samples by HPLC-ICPMS and HPLC-HG-AFS. *J. Anal. At. Spectrom.* **2015**, *30*, 2129–2183. [CrossRef]
16. Wolle, M.M.; Conklin, S.D. Speciation analysis of arsenic in seafood and seaweed: Part II—Single laboratory validation of method. *Anal. Bioanal. Chem.* **2018**, *410*, 5689–5702. [CrossRef]
17. Quiroz, W. Speciation analysis in chemistry. *Chem Texts* **2021**, *7*, 1–17. [CrossRef]
18. Narukawa, T.; Chiba, K. Heat-Assisted Aqueous Extraction of Rice Flour for Arsenic Speciation Analysis. *J. Agric. Food Chem.* **2010**, *58*, 8183–8188. [CrossRef]
19. Narukawa, T.; Suzuki, T.; Inagaki, K.; Hioki, A. Extraction techniques for arsenic species in rice flour and their speciation by HPLC-ICP-MS. *Talanta* **2014**, *130*, 213–220. [CrossRef]
20. Maher, W.A.; Eggins, S.; Krikowa, F.; Jagtap, R.; Foster, S. Measurement of As species in rice by HPLC-ICPMS after extraction with sub-critical water and hydrogen peroxide. *J. Anal. At. Spectrom.* **2017**, *32*, 1129–1134. [CrossRef]
21. Huang, J.H.; Fecher, P.; Ilgen, G.; Hu, K.N.; Yang, J. Speciation of arsenite and arsenate in rice grain—Verification of nitric acid based extraction method and mass sample survey. *Food Chem.* **2012**, *130*, 453–459. [CrossRef]
22. Heitkemper, D.T.; Vela, N.P.; Stewart, K.R.; Westphal, C.S. Determination of total and speciated arsenic in rice by ion chromatography and inductively coupled plasma mass spectrometry. *J. Anal. At. Spectrom.* **2001**, *16*, 299–306. [CrossRef]
23. Chaney, R.L.; Green, C.E.; Lehotay, S.J. Inter-laboratory validation of an inexpensive streamlined method to measure inorganic arsenic in rice grain. *Anal. Bioanal. Chem.* **2018**, *410*, 5703–5710. [CrossRef]
24. López, R.; D'Amato, R.; Trabalza-Marinucci, M.; Regni, L.; Proietti, P.; Maratta, A.; Cerutti, S.; Pacheco, P. Green and simple extraction of free seleno-amino acids from powdered and lyophilized milk samples with natural deep eutectic solvents. *Food Chem.* **2020**, *326*, 126965. [CrossRef]
25. *GB 5009.11-2014*; National Food Safety Standard-Determination of Total Arsenic and Inorganic Arsenic in Food. Food and Medical Products Administration of China: Beijing, China, 2016.
26. Narukawa, T.; Inagaki, K.; Kuroiwa, T.; Chiba, K. The extraction and speciation of arsenic in rice flour by HPLC-ICP-MS. *Talanta* **2008**, *77*, 427–432. [CrossRef]
27. Gnagnarella, P.; Salvini, S.; Parpinel, M. *Food Composition Database for Epidemiological Studies in Italy, Version 1*; IEO European Institute of Oncology: Milan, Italy, 2008; Available online: https://www.ieo.it/bda (accessed on 10 March 2022).

Article

Determination of Selenomethionine, Selenocystine, and Methylselenocysteine in Egg Sample by High Performance Liquid Chromatography—Inductively Coupled Plasma Mass Spectrometry

Yue Zhao [1,2], Min Wang [1,2,*], Mengrui Yang [1,2], Jian Zhou [1,2] and Tongtong Wang [1,2]

1. Institute of Quality Standards and Testing Technology for Agro-Products, Chinese Academy of Agricultural Sciences, Beijing 100081, China; zhaoyuecaas@163.com (Y.Z.); yangmengrui2014@163.com (M.Y.); zhoujian_8382@163.com (J.Z.); wangttong123@126.com (T.W.)
2. Key Laboratory of Agro-Product Safety and Quality, Ministry of Agriculture and Rural Affairs, Beijing 100081, China
* Correspondence: wangmin@caas.cn

Citation: Zhao, Y.; Wang, M.; Yang, M.; Zhou, J.; Wang, T. Determination of Selenomethionine, Selenocystine, and Methylselenocysteine in Egg Sample by High Performance Liquid Chromatography—Inductively Coupled Plasma Mass Spectrometry. *Separations* 2022, 9, 21. https://doi.org/10.3390/separations9020021

Academic Editors: Xianjiang Li and Rui Weng

Received: 20 December 2021
Accepted: 15 January 2022
Published: 18 January 2022

Publisher's Note: MDPI stays neutral with regard to jurisdictional claims in published maps and institutional affiliations.

Copyright: © 2022 by the authors. Licensee MDPI, Basel, Switzerland. This article is an open access article distributed under the terms and conditions of the Creative Commons Attribution (CC BY) license (https://creativecommons.org/licenses/by/4.0/).

Abstract: The deficiency of selenium in dietary is recognized as a global problem. Eggs, as one of the most widely consumed food products, were readily enriched with selenium and became an important intake source of selenium for humans. In order to better understand the speciation and bioaccessibility of selenium in eggs, a simple and reliable approach that could be easily used in a routine laboratory was attempted to develop for analyzing selenium species. Three of organic selenium species (selenocystine, methylselenocysteine, and selenomethionine) in liquid whole egg were completely released by enzymatic hydrolysis and detected by high performance liquid chromatography in combination with inductively coupled plasma mass spectrometry (HPLC-ICP-MS). All the parameters in enzymatic hydrolysis and separation procedures were optimized. The effect of matrix in analysis was critically evaluated by standard addition calibrations and external calibrations. Under the optimal conditions, the spike recoveries of selenium species at 0.1–0.4 $\mu g\ g^{-1}$ spike levels all exceeded 80%. This method was successfully applied to the determination of selenium species in fresh egg and cooked eggs.

Keywords: selenium; speciation; enzymatic hydrolysis; HPLC-ICP-MS; egg

1. Introduction

Selenium (Se), as a crucial microelement, is an indispensable nutrient for normal activity of human body [1]. Previous studies have reported the role of Se in protecting human body from acne, psoriasis, multiple sclerosis, colorectal cancer, cervical dysplasia, esophageal cancer, and stomach cancer [2]. The deficiency of Se can induce a number of degenerative diseases, such as endemic Kashin–Beck disease [3]. Human health may be improved, if intake of Se is increased in a reasonable range. And the main and safe source of selenium supplementation for humans is diet. Thus, Se-enriched foods which have beneficial functions to our health have been emerging in recent years.

Eggs, which have been regarded as a source of bioactive compounds of animal origin for many years, are considered as an important constituent of diet due to their nutritional value, such as enrichment in proteins, vitamins, and minerals [4,5]. Se content in eggs can easily be manipulated to give increased levels [6]. Therefore, Se-enriched eggs are currently sold in many countries and have become the main source of Se supplementation for humans. Nowadays Se-enriched eggs on the market are usually labeled only by content of total selenium. But the bioavailability, metabolic pathway and physiological functions of Se were not only determined by the intake level, but also depend on the chemical forms of element [7–9]. Therefore, except for determining the total Se content, the composition of Se species in food should be fully considered when evaluating the Se-enriched food quality.

Researchers have spent tremendous effort to perform analysis of Se and its speciation in eggs. To date, a series of techniques (i.e., spectrofluorometry, atomic spectrometry) was reported for the analysis of total Se in eggs [10–12]. However, speciation studies of Se in eggs are still scarce in literature. Lipiec et al. attempted to use high-performance liquid chromatography combined with inductively coupled plasma mass spectrometry (HPLC-ICP-MS) simultaneous determining selenomethionine (SeMet), selenocysteine (SeCys), and inorganic Se(IV) in egg white lyophilisate powder and yolk protein precipitate which were defatted by cyclohexan. The detection limits were 60, 3, and 10 µg kg^{-1} (dry weight) for SeCys, Se(IV), and SeMet, respectively, and the precision was 5–10% [13]. Sun and Feng developed a method using HG-AFS by depositing albumen with trichloroacetic acid to determination organic and inorganic Se in fresh Se-enriched eggs [14]. Up to now, there were few reports about the analysis method of Se species in fresh liquid whole egg.

It was well known that most of Se was believed to be incorporated into proteins through biotransformation in animal. And organic Se species (SeMet, selenocystine (SeCys$_2$), and methylselenocysteine (MeSeCys)) were more effective in alleviating deficiency symptoms [15,16]. In this work, the existence of SeCys$_2$, SeMet, and MeSeCys in eggs was mainly concerned. HPLC-ICP-MS is one of the most commonly used methods due to its high sensitivity towards Se and tolerance to the sample matrix [17–19]. Therefore, a simple procedure applying in routine laboratory to determine the concentrations of individual seleno-amino acides (SeCys$_2$, SeMet, and MeSeCys) in fresh liquid whole eggs was attempted to be developed using HPLC-ICP-MS. Enzymatic hydrolysis was used for releasing the Se-amino acid from proteins. Particular attention was paid to the optimization of enzymatic hydrolysis procedure, and matrix effect in analysis was also critically evaluated. Finally, this method was applied to the Se speciation of fresh egg and cooked egg.

2. Materials and Methods

2.1. Reagents and Solutions

Chemicals/solvents used in this study were analytical reagent or higher. Ultrapure water (Millipore, Burlington, MA, USA) was used in all samples and standard preparations. Seleno-DL-cystine (SeCys$_2$), Methylselenocysteine (MeSeCys, 95%), Seleno-L-methionine (SeMet, 98%), sodium selenite (Se(IV), 99%), and sodium selenate (Se(VI)) were purchased from Sigma-Aldrich (Shanghai, China). The enzymes: Protease XIV (from Streptomyces griseus), Protease XIII (from Aspergillus saitoi), Protease XXIV (from bacterial), Pronase K (from Tritirachium album), Protease VIII (from Bacillus licheniformis), Pronase (from Streptomyces griseus), Papain (from papaya latex), Trypsin (from porcine pancreas), Pepsin (from porcine gastric mucosa) and Lipase (from porcine pancreas) were purchased from Sigma-Aldrich. Ammonium acetate (HPLC grade) was purchased from Fisher Scientific (Shanghai, China). Methanol (HPLC grade) was purchased from Honeywell Burdick & Jackson (Muskegon, MI, USA). Tris(hydroxymethyl)amino methane hydrochloride and Tetrabutylammonium hydroxide (TBAH 40 wt.% solution in methanol) was obtained from Shanghai Macklin Biochemical Co., Ltd. (Shanghai, China). NaH$_2$PO$_4$·2H$_2$O, Na$_2$HPO$_4$·12H$_2$O, NaHCO$_3$, and NH$_4$HCO$_3$ were purchased from Beijing Chemical Works (Beijing, China).

The certain amount standards of five Se species were each accurately weighed and transferred into individual 50 mL brown glass volumetric flasks. They were dissolved with ultrapure water to produce five Se species standard stock solutions of 50 mg L^{-1} (as Se element). The stock solutions were stored at 4 °C in dark and used within one month. A mixed standard working solution (5 mg L^{-1}) containing equal concentrations of five Se species was prepared from these individual stock solutions by the gradient dilution method and serially diluted to make 2, 5, 10, 30, 50, and 100 µg L^{-1} standards.

2.2. Sample Collection and Preparation

The liquid whole egg sample that used for all of optimizations was made up of egg white and egg yolk. In order to ensure the homogeneity of liquid whole egg sample, several eggs of one brand were sufficiently mixed in a beaker using an electric eggbeater. Then homogenized liquid whole egg was divided into PTEF sample vials (50 g per bottle) under nitrogen gas protection, and stored in freezer ($-70\ °C$). Before sample pretreatment, the liquid whole egg sample was taken from freezer ($-70\ °C$) and thawed at room temperature.

The Se-enriched eggs of one brand used in cooking processing were obtained in local store. The boiled egg sample was prepared by fresh egg immersed in 500 mL tap water, then boiled for 15 min. Then the samples were cooled to room temperature, peeled and mixed with a mixer. The poached egg sample was prepared by peeling fresh egg immersed in 500 mL of boiling tap water for 10 min. Then the samples were drained, cooled to room temperature and mixed with a mixer. The egg flower soup was prepared by mixed fresh egg liquid poured in 500 mL of boiling tap water for 2 min. Then the solid was drained, cooled to room temperature and mixed with a mixer. The steamed egg custard was prepared by the egg mixture (fresh egg liquid:water = 1:1.5) steamed for 10 min in 300 mL boiled tap water. Then the solid was cooled to room temperature and mixed with a mixer. The experimental procedures were carried out in triplicate. Selenium was not detected in tap water used for cooking.

2.3. Instrumental Analysis

An Agilent 7900 ICP-MS (Agilent Technologies, Santa Clara, CA, USA) was used for establishing the method. The optimized settings of ICP-MS were as follows: sampling depth, 8 mm; RF power, 1550 W; temperature of the atomizer chamber, $2\ °C$; carrier gas flow rate, 0.85 L min^{-1}; make-up gas, 0.20 L min^{-1}; He gas flow rate, 3.8 mL min^{-1}; acquisition mode, time resolved analysis (TRA); monitored isotope, ^{78}Se and ^{80}Se; integration time, 0.5 s.

For HPLC, an Agilent 1260 Infinity II LC system (Agilent Technologies, Santa Clara, CA, USA) with a binary pump and autosampler were used. The column outlet of HPLC system was connected to micro-Mist nebulizer using PEEK tubing (0.25 mm i.d. × 104 cm length). The mobile phase in ion-pairing reversed-phase chromatography (Columns: Agilent Extend C18 (4.6 mm × 250 mm, 5 μm), Agilent Eclipse XDB C18 (4.6 mm × 250 mm, 5 μm), Agilent Eclipse Plus C18 (4.6 mm × 250 mm, 5 μm), and Agilent StableBond (4.6 mm × 250 mm, 5 μm)) was composed of 0.5 mM TBAH, 10 mM ammonium acetate, and 2% methanol (v/v). The pH of the solution mentioned above was adjusted with adding acetic acid to 5.5. The mobile phase in anion-exchange chromatography (Column: Hamilton PRP X-100 (4.1 mm × 250 mm, 10 μm)) was composed of 5 mM citric acid (pH = 5.0). Before use, they were filtered through 0.45 μm filters and degassed by ultrasound. The flow rate was set to 1 mL min^{-1}. 50 μL of sample was injected using an auto-sampler device. The samples were separated at room temperature ($25\ °C$).

2.4. Microwave Digestion for Total Se Determination

A sample (2 g whole egg sample or 2 mL supernatant of extracts) was digested with a mixture of 5 mL of HNO$_3$ and 2 mL of H$_2$O$_2$ using closed-vessel microwave digestion system (TOPEX, PreeKem, Shanghai, China). The digestion conditions were as follows: 3 min at $80\ °C$, 3 min at $100\ °C$, 3 min at $130\ °C$, 3 min at $160\ °C$, and 25 min at $190\ °C$. The volume of solution was reduced below 1 mL using an electric evaporation block at $120\ °C$. After cooling down, the digest was transferred to volumetric flask, diluted to 25 mL with ultrapure water, and filtered through a 0.45 μm filter. The procedure was performed in triplicate and blank tests were also carried out. The accuracy and precision of this method were validated against reference material GBW 10018 (National Research Center, Beijing, China).

2.5. Extraction of Se Species from Egg Using Different Procedures

Five extraction procedures, water extraction, acid extraction (0.1 M HCl), basic extraction (0.1 M NaOH), buffer extraction (25 mM ammonium acetate buffer containing 5% of methanol (v/v)), and enzymatic hydrolysis were studied. Except for enzymatic hydrolysis, four extraction procedures were carried out as follows: 10 mL of corresponding solution was added to 2.5 g of homogenized liquid whole egg in 50 mL centrifuge tube. Samples were sonicated for 30 min at room temperature. Then, the mixtures were thoroughly homogenized by shaking in water bath shaker (165 rpm, 37 °C) for 18 h. After extraction process, samples were centrifuged for 5 min at 10,000 rpm. The supernatant was transferred to other centrifuge tube and stored at −70 °C until analysis. Measurements of total Se (for extraction yield) in extracts were performed by microwave digestion inductively coupled plasma mass spectrometry method. For Se species analysis, the extracts were filtered using 10.0 kDa ultrafiltration membranes (Merck Millipore Ltd., Tullagreen, Carrigtwohill, County Cork, Ireland). In the batch of enzymatic hydrolysis, 10 mL of ultrapure water were added to 2.5 g of liquid whole egg in 50 mL centrifuge tube with sonicating for 30 min. Then 50 mg protease XIV was added to the centrifuge tube. Other details of the experiment were the same as that shown above. The procedure was performed in triplicate and blank tests were also carried out.

2.6. Optimization of the Enzymatic Hydrolysis Procedure

In this section, the parameters of enzymatic hydrolysis procedures (i.e., extraction solution, types and amount of enzyme) were optimized. The five extraction solutions (ultrapure water, Tris-HCl (100 mM, pH = 7.5), PBS (100 mM, pH = 7.5), $NaHCO_3$ (100 mM, pH = 8), and NH_4HCO_3 (100 mM, pH = 8) were studied in the system with 50 mg protease XIV. The extraction procedures of different enzyme type and enzyme usage were investigated under Tris-HCl (100 mM, pH = 7.5) buffer solution. Other details of experiment (i.e., experimental temperature, time and so on) were the same as that listed in the previous section. Measurements of Se species in the extracts were performed by HPLC-ICP-MS. Meanwhile, spike recovery test was also applied. A mixture of $SeCys_2$, MeSeCys, and SeMet (0.2 µg g^{-1} (in egg)) was added to the sample. The spiked sample was mixed thoroughly in a vortex mixter (IKA, Staufen, Germany) for several minutes and then placed for 2 h in order to make the added Se species mixed well with the sample. Then the pretreatment procedure of spiked sample was the same as unspiked sample. Optimum conditions of enzymatic hydrolysis were selected by comprehensive consideration the concentration of Se species extracted from unspiked egg and the spike recovery of spiked sample.

2.7. Method Validation

2.7.1. Linearity

The linearity of the method was evaluated within the standard concentration range for liquid whole egg. Linear regression (weighted $1/x^2$) was used to produce the best fit for the concentration/peak area ratio relationship for $SeCys_2$, MeSeCys, SeMet, Se(IV), and Se(VI) in ultrapure water. The limit of detection (LOD) and quantitation (LOQ) was defined as the concentration corresponding to 3 times and 10 times signal to noise (S/N), respectively.

2.7.2. Recovery Assay

The accuracy and precision of the method were characterized by spiked liquid whole egg. The calculated volumes of mixed standard solutions of Se species were added to samples to generate fortified samples with three different spiked concentrations (0.1, 0.2 and 0.4 µg g^{-1}). Subsequently, enzymatic hydrolysis and HPLC-ICP-MS detection were performed using the optimal conditions. Five replicate treatments for each spike concentration were carried out. The quantitative results by external calibration method and standard addition calibration method were used for calculating the spike recovery, respectively.

2.7.3. Matrix Effect

Matrix effect refers to the matrix often significantly interfering with the analyte analysis process. It is an important parameter that affects the selectivity and sensitivity of quantitative analysis, and the matrix effect is calculated with the following Equation (1):

$$ME\,(\%) = \left(\frac{k_1}{k_2} - 1\right) \times 100 \tag{1}$$

where ME (%) is matrix effect; k_1 is the slope of matrix standard curve; k_2 is the slope of solvent standard.

The matrix effects can be divided into three regions according to the value of ME (%). No matrix effect is considered when ME is in the range of −20–20%. When ME is between 20 and 50% or between −50 and −20%, there is a medium matrix effect, while a strong matrix effect is less than −50% or more than 50%. Positive and negative values mean signal enhancement and suppressions induced by matrix, respectively [20].

3. Results and Discussion

3.1. Evaluation of Different Sample Extraction Procedures

The aim of this study was to establish a method that could both efficiently extract of Se species and be able to keep the original Se speciation information in egg. For this purpose, several sample preparation methods (i.e., extraction with nonbuffered and buffered solution, enzymatic hydrolysis) were selected for Se speciation of liquid whole egg sample [2,16–19]. The concentration of total Se in supernatants under different preparation procedure was determined by microwave-ICP-MS. The extraction efficiency of Se was calculated by following equation: extraction efficiency (%) = amount of Se detected in supernatant/amount of Se detected in liquid whole egg × 100.

The extraction efficiencies of Se with different pretreatment procedure (extraction with ultrapure water, HCl, NaOH, ammonium acetate buffer, or protease XIV) were shown in Figure 1. Compared with the pretreatment method of ultrapure water and buffer solution, the concentrations of Se extracted with HCl, NaOH and enzyme were higher, accounting for more than 60% of the total Se content (218.40 ± 21.10 µg kg^{-1}) of egg (Figure 1). Then, the supernatant was analyzed by HPLC-ICP-MS for Se species (Figures 1 and S1). As shown in Figure S1, no peak was observed in the chromatogram of supernatant extracted with HCl, and two unknown peaks were observed in that with NaOH. However, large amounts of target organic Se species (SeCys$_2$ and SeMet) were observed in the system with enzymatic hydrolysis. The concentrations of SeCys$_2$ and SeMet were 81.80 ± 0.53 µg kg^{-1} and 6.96 ± 0.20 µg kg^{-1} (as Se), respectively. According to reports, the intake of Se converted by hen organism was bonded to protein and present mainly in organic forms in eggs [13,14]. The results of Se species extracted from egg by enzymatic hydrolysis were consistent with this conclusion. Enzymatic hydrolysis method had high extraction efficiency and could guarantee the integrity of Se species in samples [21]. Therefore, enzymatic hydrolysis was selected as the pretreatment method for Se speciation of egg, and the optimization of enzymatic hydrolysis was investigated below.

3.2. Optimization of Enzymatic Hydrolysis Parameters

3.2.1. Extraction Solution

In the literature, different extraction solutions have been applied to a variety of food matrices to achieve high extraction efficiency of Se species from samples [7,22]. In this work, the study of extraction solutions was carried out both by comparing the extracted original concentration of Se species from eggs and the results of spike recovery test. The experiment results were listed in Table 1.

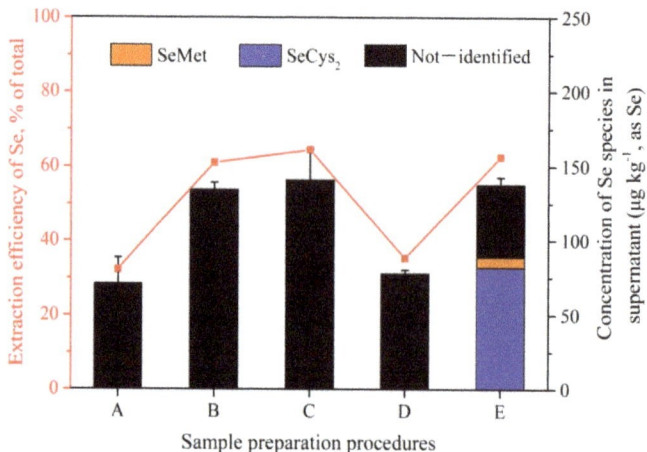

Figure 1. Extraction efficiency of Se and concentration of Se species in supernatant with different preparation procedures: (A) Extraction with ultrapure water; (B) Extraction with HCl (0.1 M); (C) Extraction with NaOH (0.1 M); (D) Extraction with 20 mM ammonium acetate solution containing 5% of methanol (v/v); (E) Hydrolysis with protease XIV. The value of Se in supernatant minus the observed Se species was defined as the "not-identified" Se species.

Table 1. Concentration and spike recoveries of SeCys$_2$, MeSeCys, and SeMet in eggs using different extraction solution (Results expressed as the mean ± standard deviation (SD)).

Extraction Solution	[Se] Species (µg kg^{-1})				
	SeCys$_2$	MeSeCys	SeMet	[Se]$_{sum\ of\ species}$	R(%) [a]
ultrapure water	81.80 ± 0.82	ND [b]	6.96 ± 0.20	91.58 ± 1.24	41.84 ± 0.57
Tris-HCl	83.72 ± 0.36	ND	8.00 ± 0.10	92.93 ± 0.25	42.46 ± 0.12
PBS	45.01 ± 1.10	ND	9.68 ± 1.36	56.58 ± 2.34	25.85 ± 1.07
NaHCO$_3$	59.70 ± 8.02	ND	8.96 ± 1.79	69.92 ± 6.11	31.95 ± 2.79
NH$_4$HCO$_3$	59.37 ± 0.95	ND	9.31 ± 0.31	69.70 ± 0.06	31.85 ± 0.03
	Recoveries (%) of Se Species in the Spiked Sample				
Extraction Solution	SeCys$_2$	MeSeCys		SeMet	
ultrapure water	69.95 ± 5.33	73.38 ± 2.09		78.98 ± 0.52	
Tris-HCl	67.83 ± 1.05	69.26 ± 1.54		80.37 ± 0.17	
PBS	67.18 ± 0.59	67.13 ± 1.66		80.23 ± 2.71	
NaHCO$_3$	58.95 ± 1.76	56.13 ± 2.87		79.15 ± 0.96	
NH$_4$HCO$_3$	58.47 ± 0.18	68.05 ± 0.39		79.48 ± 1.48	

[a] R(%) calculated as sum of Se species related to total Se found in egg. [b] ND: not detected.

As shown in Table 1, the extracted amounts of three Se species were affected largely by extraction solution. The sum of extracted three Se species using Tris-HCl buffer solution was largest among five extraction solutions, and the value was 92.93 ± 0.25 µg kg^{-1}, which accounted for 42.5 ± 0.1% of the total Se found in eggs. Meanwhile, the sum of three Se species extracted by PBS buffer solution was the lowest (45.01 ± 1.10 µg kg^{-1}). But interestingly, the spike recoveries of SeCys$_2$, MeSeCys, and SeMet were influenced by extraction solution slightly. The average spike recoveries of SeCys$_2$, MeSeCys, and SeMet with different extraction solutions were 64.5 ± 8.3%, 66.9 ± 9.6%, and 79.6 ± 0.8%, respectively. The Se species added to samples were water-soluble amino acids (SeCys$_2$, MeSeCys, and SeMet), which were different from that bound to actual samples. Therefore, activity of enzyme influencing by extraction solution affected slightly on the spike recoveries of three Se species. Considering extracted concentrations from egg and spike recoveries of Se species in different extraction solution system overall, Tris-HCl buffer solution (100 mM, pH 7.5) was chosen as the optimal extraction solution in this study.

3.2.2. Protease

Recently, the common proteases developed in market are mainly animal-derived proteases, plant-derived proteases and microbial-derived proteases. According to previous studies, this work selected animal-derived proteases (pepsin and trypsin), plant-derived proteases (papain), and microbial-derived proteases (protease XIII, protease XXIV, protease K, protease VIII, protease XIV, and pronase) to evaluate the extraction of Se species in liquid whole egg [2,21–23]. The spike recoveries of three Se species in system with different enzymes were shown in Figure 2.

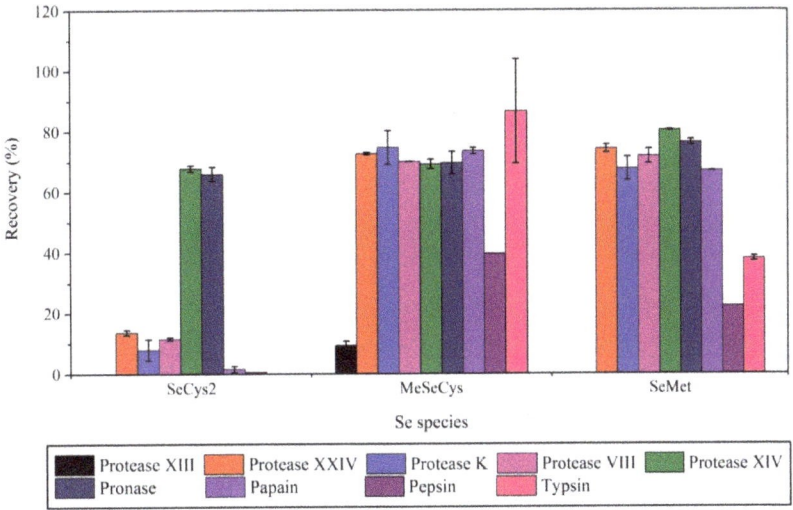

Figure 2. Spike recoveries of three Se species with different enzymes.

As shown in Figure 2, the spike recoveries of three selenium species in system with animal-derived protease were poor. In trypsin system, SeCys$_2$ was almost not extracted (the spike recovery was less than 10%), while the spike recovery of MeSeCys was 86.6%. In pepsin system, the spike recoveries of three Se species were all below 40%. In plant-derived protease system (i.e., papain), SeCys$_2$ was not extracted while the spike recoveries of other two Se species were 73.6% for MeSeCys and 67.0% for SeMet. Among the microbial-derived proteases, protease XIV and pronase were suitable for the extraction of three Se species from eggs. In the enzymatic hydrolysis procedure with protease XIV or pronase, the spike recoveries of three Se species were both above 67.8%.

It was worth noting that the spike recoveries of three Se species in system with protease XIV or pronase were similar, which might be attributed to the same source of two enzymes. The highest spike recovery of Se from antarctic krill samples was also achieved by pronase E [21]. In addition, the concentration of Se species (85.74 ± 1.36 µg kg^{-1} for SeCys$_2$, and 8.23 ± 0.25 µg kg^{-1} for SeMet) extracted from unspiked sample by enzymatic hydrolysis system with pronase was largest. Finally, enzymatic hydrolysis procedure with pronase was chosen for the quantitative Se species extraction of liquid whole eggs in this work.

3.2.3. Pronase Usage

Optimization of enzyme usage was performed to find out the optimal Sample/Enzyme ratio in extraction step. Three Se species in egg sample were extracted using 10.0 mL of 100 mM Tris HCl (pH 7.5) containing varying amounts of pronase (10, 25, 50, 75, and 100 mg) for 18 h. As shown in Figure 3, the spike recoveries of SeCys$_2$ and SeMet increased with increasing the amount of pronase (10–75 mg). When the amount of pronase reached to 100 mg, the spike recoveries of SeCys$_2$ and SeMet were decreased compared to those of 75 mg pronase system. And the maximum spike recoveries were obtained under the system containing 75 mg of pronase. On the contrary, the spike recovery of MeSeCys did not change obviously with the amount of pronase increasing.

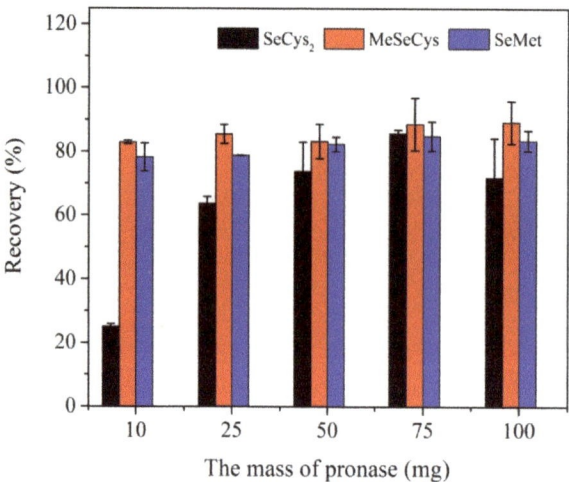

Figure 3. Spike recoveries of three Se species with different pronase usage.

The Se species in unspiked egg sample extracted by enzymatic hydrolysis with different pronase usage were also analyzed in this work. The experiment results showed that extracted concentration of SeCys$_2$ was significantly affected by the usage of pronase (10 mg: 22.76 ± 2.36 µg kg^{-1} (as Se); 75 mg: 87.94 ± 3.07 µg kg^{-1} (as Se)). When the usage of pronase in system was small, the enzymatic hydrolysis of protein in eggs was incomplete. With the usage of pronase increasing from 75 mg to 100 mg, extracted concentration of SeCys$_2$ was almost unchanged (75 mg: 87.94 ± 3.07 µg kg^{-1} (as Se); 100 mg: 89.16 ± 3.90 µg kg^{-1} (as Se)). This result was not quite the same as spike recoveries of SeCys$_2$ (75 mg: 85.8 ± 1.1%; 100 mg: 71.9 ± 12.4%). In addition, the extraction concentrations of other two Se species were affected slightly by the usage of pronase in system. Finally, 75 mg of pronase was selected as the optimal usage in further experiments for 2.5 g of liquid whole egg sample. The optimal enzyme/sample ratio of eggs (3:100) in this study was higher than that in enzymatic hydrolysis of chicken breast sample (1:10) [2]. This may be due to higher moisture content (70–80%) of liquid whole egg than that of chicken breast lyophilized powder sample.

3.2.4. Lipase Usage

As is well known, eggs are not only rich in protein (14–16%) but also high in fat (10–12%). The analysis of Se species was interfered by fat present in sample. Fat cannot be hydrolyzed by protease; therefore, the introduction of lipase in pretreatment process was necessary. In this work, Se in egg sample was extracted using 10.0 mL of 100 mM Tris-HCl (pH 7.5) containing 75 mg of pronase and varying amounts of lipase (0, 10, 25, 50, 75, and 100 mg) for 18 h. And the results were showed in Figure 4.

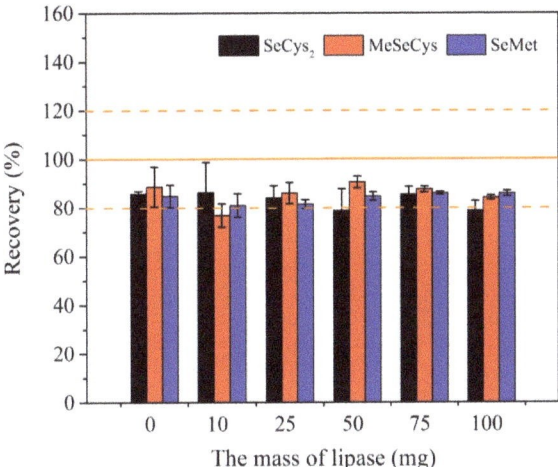

Figure 4. Spike recoveries of three Se species with different lipase usage. Solid line represents the spike recovery value of 100%. Dash lines represent spike recovery range between 80% and 120%.

As shown in Figure 4, the spike recoveries of Se species were affected by lipase usage slightly. The average spike recoveries of Se species were 83.2 ± 3.5%, 85.7 ± 4.8%, and 84.0 ± 2.2% for SeCys$_2$, MeSeCys, and SeMet, respectively. The spike recoveries of three Se species were optimum in the system with 75 mg of lipase (Recovery > 80% and RSD < 5%), as shown in Figure 4. And in this condition, more Se species extracted from unspiked sample were obtained. The concentration of SeCys$_2$ was 90.79 ± 5.29 µg kg^{-1}, and the concentration of SeMet was 17.14 ± 0.95 µg kg^{-1}. A mixture of 75 mg pronase and 75 mg lipase was selected as the optimal usage of enzyme in further experiments for 2.5 g of liquid whole egg sample. With the addition of lipase, the supernatant of samples became clearer after enzymatic hydrolysis procedure. It was indicated that egg sample underwent a higher extent of enzymatic hydrolysis. The use of lipase made the supernatant easier to filter through 10.0 kDa ultrafiltration membranes.

3.3. Chromatographic Separation

Ion-exchange chromatography and ion-pairing reversed-phase chromatography have been widely used for the separation of Se compounds. In this study, two types of columns as mentioned previously were selected to evaluate their performances in Se species separation. For the ion-pairing reversed-phase chromatography systems, four typical reversed C18 columns, namely, Agilent StableBond (SB-Aq) (4.6 mm × 250 mm, 5 µm), Agilent Extend C18 (4.6 mm × 250 mm, 5 µm), Agilent Eclipse XDB C18 (4.6 mm × 250 mm, 5 µm) and Agilent Eclipse Plus C18 (4.6 mm × 250 mm, 5 µm) were tested. The compositions of mobile phases were given in the section of instrumental analysis.

As shown in Figure 5, five Se species can be well separated on Hamilton PRP X-100 column with citric acid as the eluent within 20 min. The retention times of SeCys$_2$, MeSeCys, Se(IV), SeMet, and Se(VI) were 2.23, 2.89, 3.82, 5.51, and 16.67 min, respectively. The peak intensity of SeCys$_2$ (30,615 CPS of peak height) was highest among five Se species. In ion-pairing reversed-phase chromatography, excellent sensitivities of all tested Se species were obtained due to the methanol in mobile phase boost signal intensity for Se by ICP-MS. The peak intensities for five Se species obtained using ion-pairing reversed-phase chromatography were notably 3 times higher than on Hamilton PRP X-100 column (Figure 5).

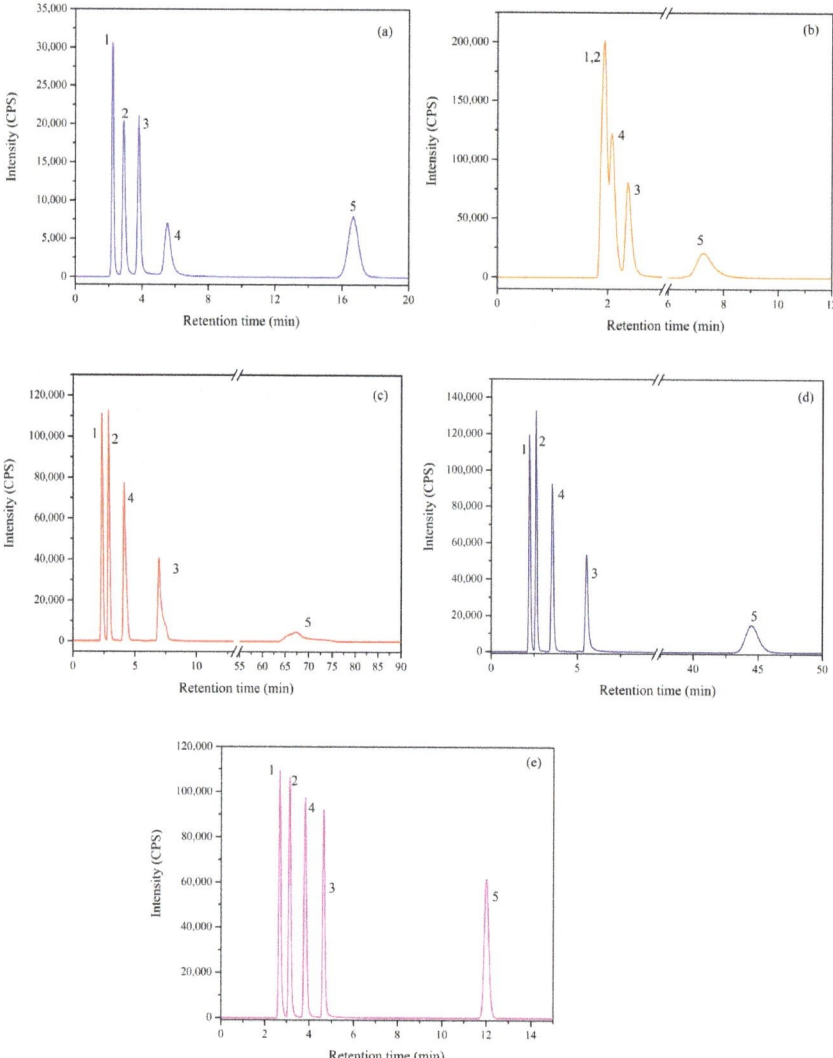

Figure 5. Chromatograms of a mixture of Se species at 500 µg L^{-1} (as Se) in solution on anion-exchange column ((**a**) Hamilton PRP X-100 (4.1 mm × 250 mm, 10 µm)) and reversed-phase columns ((**b**) Agilent Extend C18 (4.6 mm × 250 mm, 5 µm)); (**c**) Agilent Eclipse XDB C18 (4.6 mm × 250 mm, 5 µm); (**d**) Agilent Eclipse Plus C18 (4.6 mm × 250 mm, 5 µm); (**e**) Agilent SB-Aq (4.6 mm × 250 mm, 5 µm)). The peaks of five Se species in chromatograms were (1) SeCys$_2$, (2) MeSeCys, (3) Se(IV), (4) SeMet, and (5) Se(VI).

Although Extend C18, Eclipse XDB C18, Eclipse Plus C18 and SB-Aq columns have the same retention mechanism, the retention behavior of five Se species were markedly different on these columns. Organic Se species were easily separated on Eclipse XDB C18, Eclipse Plus C18, and SB-Aq columns within 6 min. However, co-elution of three organic Se species was observed on Extend C18 column. Compared with SB-Aq column, the peak shape and intensities for inorganic Se species obtained from Extend C18, Eclipse XDB C18, and Eclipse Plus C18 column were poor. In addition, the retention time of Se(VI) was 67.49 min in Eclipse XDB C18 columns. Overall, for simultaneous determination of organic and inorganic Se compounds, SB-Aq column could provide the least peak broadening and symmetry, as well as closely eluting peaks and satisfactory resolution of both compounds (R > 1.5). Therefore, Agilent SB-Aq column was selected for the further analysis of Se compounds in egg samples.

3.4. Analytical Performance of the Method

The analytical performance data (i.e., calibration curves, linear correlation coefficients, and so on) under optimal instrument conditions were summarized in Table S1. The working curve of Se species was linear (R > 0.999) in concentration range of 2–100 µg L^{-1}. The LOQs of HPLC–ICP-MS were calculated as 1.86 µg L^{-1} for SeCys$_2$, 1.62 µg L^{-1} for MeSeCys, 1.82 µg L^{-1} for SeMet, 2.37 µg L^{-1} for Se(IV), and 1.98 µg L^{-1} for Se(VI), respectively. In the enzymatic extracts of eggs, except Se species, amino acids and their salts, chlorides and other substances might exist. To determine Se species accurately, matrix effects were evaluated by performing standard addition calibrations and external calibrations. The calculated MEs using Equation (1) were −13% for SeCys$_2$, 8% for MeSeCys, and 9% for SeMet. These MEs were between −20% and 20%, indicating that there was no matrix effect [20]. In addition, the values of spike recovery were tested by F-test. At a confidence level of 95%, no significant differences were found. And recoveries of three Se species were in the range from 85.5% to 98.2% (Table 2), within the limit of 80–120%, showing that the matrix effects could be excluded. Therefore, the method was capable of analyzing three seleno-amino acid in liquid whole egg satisfactorily.

Table 2. Spike recoveries of Se species quantified by standard addition calibration and external calibration (n = 5).

Se Species	Spiked Level (µg g^{-1})	Recovery (%) Standard Addition Calibration		Recovery (%) External Calibration	
SeCys$_2$	0.1	96.7	2.9	98.2	3.7
	0.2	86.9	4.0	85.5	3.2
	0.4	91.6	0.3	90.6	2.1
MeSeCys	0.1	96.5	2.0	90.2	4.5
	0.2	89.6	7.3	87.7	1.2
	0.4	92.0	3.8	91.3	0.6
SeMet	0.1	86.9	1.8	87.9	0.7
	0.2	88.5	4.1	86.0	0.6
	0.4	87.8	0.2	88.6	0.9

3.5. Determination of Se Species in Real Egg Samples

3.5.1. Se Speciation of Fresh Egg

The developed method was employed for the determination of Se species in different brands of regular eggs and Se-enriched eggs, which were purchased from local stores. As shown in Table 3, SeCys$_2$ and SeMet were the dominant Se species in egg, and they added up to about 59 ± 4% of total Se. There was no significant difference in the content of SeCys$_2$ detected in regular and Se-enriched eggs, but the content of total Se and SeMet had significant difference. The concentration of SeMet increased significantly in Se-enriched eggs. No MeSeCys was identified in all tested egg samples. It was noting that other Se species were not observed in all tested eggs (Figure S2), and the unknown Se species were accounted for 40% of total Se content. Except for SeCys$_2$, MeSeCys, and SeMet, Se element could present as other seleno-amino acids (i.e., selenourea, selenoethionine, selenomethionine-Se-oxide, and so on) in organisms [7,23,24]. In further studies, more methods and techniques (i.e., UHPLC-ESI-QqQ-MS/MS or UHPLC-ESI-Orbitrap-MS/MS) should be attempted for the identification of unknown Se species in eggs.

Table 3. Results of Se speciation analysis in different egg samples [a].

Sample [b]	SeCys$_2$	MeSeCys	SeMet	[Se]$_{sum\ of\ species}$	Total Se
1	129.25 ± 1.75	ND [c]	13.87 ± 2.50	143.12 ± 0.75	238.54 ± 12.50
2	117.15 ± 1.98	ND	18.56 ± 1.53	135.71 ± 0.45	218.89 ± 10.50
3	92.22 ± 1.50	ND	18.38 ± 1.22	110.44 ± 1.97	205.39 ± 19.76
4	120.09 ± 2.36	ND	20.39 ± 2.92	140.48 ± 5.16	259.84 ± 15.75
5	83.71 ± 3.50	ND	120.29 ± 1.54	204.0 ± 1.96	347.60 ± 11.50
6	78.65 ± 1.78	ND	108.46 ± 3.50	187.11 ± 1.72	313.85 ± 21.50
7	97.46 ± 4.17	ND	240.22 ± 0.38	335.58 ± 5.42	547.15 ± 19.75
8	98.69 ± 13.28	ND	79.56 ± 20.10	178.25 ± 6.83	284.03 ± 20.71
9	103.68 ± 5.83	ND	108.67 ± 17.36	212.35 ± 11.64	335.66 ± 22.40
10	118.62 ± 0.26	ND	148.36 ± 5.41	266.97 ± 0.52	509.56 ± 8.44

[a] The contents of all Se species are in units of μg kg^{-1} (as Se). Results expressed as the mean ± standard deviation (SD); n = 3 individual eggs for each brands. [b] 1–4: regular egg; 5–10: Se-enriched egg. [c] ND: not detected.

3.5.2. Se Speciation in Egg Cooking Procedures

Eggs are usually consumed using various cooking methods. Several studies have shown that food processing can modify the species distribution [16,25]. Therefore, the influence of cooking process on content of total Se and Se species in Se-enriched eggs was evaluated. Once cooked, the samples were submitted to enzymatic hydrolysis by using the optimal conditions applied to fresh egg. The extracts were analyzed by HPLC-ICP-MS. After the cooking processing, new Se species was not observed in egg samples. The percentage of SeCys$_2$ was 29.3% for boiled egg, 33.9% for poached egg, 30.0% for egg flower soup, and 45.0% for steamed egg custard. Meanwhile, the percentage of SeMet was 25.3% for boiled egg, 25.7% for poached egg, 19.6% for egg flower soup, and 32.3% for steamed egg custard. It was worth noting that the percentages of SeCys$_2$ and SeMet in steamed egg custard were the highest. The main reason of seleno-amino acid transformation has been attributed to the high temperature applied in cooking that may change the chemical distribution of Se species [16].

As shown in Table 4, when eggs were cooked by boiling, a larger amount of total Se was released from egg to water without protection by egg shell. The amount of total Se in aqueous of egg flower soup was as high as 42.8% of total Se in whole egg. Meanwhile, SeCys$_2$ and SeMet were detected in aqueous. Therefore, when Se-enriched eggs used for cooking egg flower soup, it was recommended that the aqueous and solid were eaten together to ensure the full uptake of Se in eggs.

Table 4. Concentration of Se species in cooked eggs [a].

Processing	Boiled Egg	Poached Egg	Egg Flower Soup	Steamed Egg Custard
SeCys$_2$	91.03 ± 2.47	105.15 ± 5.73	94.67 ± 6.72 (32.92 ± 3.65)	170.38 ± 26.63
SeMet	78.52 ± 4.35	79.68 ± 7.67	61.75 ± 4.65 (15.99 ± 4.77)	122.23 ± 23.68
Total Se	310.66 ± 8.86 (4.82 ± 1.03)	309.98 ± 17.60 (24.03 ± 13.40)	315.26 ± 5.13 (134.92 ± 10.57)	378.29 ± 25.59

[a] The contents of all Se species are in units of μg kg^{-1} (as Se). Results expressed as the mean ± standard deviation (SD, n = 3). The value in bracket indicates the content of total Se or Se species released into the water from the egg after cooking processing.

4. Conclusions

A simple and accuracy quantitative assay based on enzymatic hydrolysis pretreatment and HPLC-ICP-MS had been developed for Se speciation in eggs. Parameters for efficient pretreatment procedures and HPLC-ICP-MS analysis were carefully optimized. Agilent StableBond column was selected to separate Se species and the working curve was linear (R > 0.999) in concentration range of 2–100 μg L^{-1}. Se species in egg was completely released into free forms by enzymatic hydrolysis with pre-treating with 100 mM Tris-HCl (pH = 7.5) at ultrasound bath for 30 min and hydrolyzing by the combination of 75 mg pronase and 75 mg lipase at water bath shaker (37 °C, 165 rpm) for 18 h. The effect of matrix in analysis was critically evaluated by standard addition calibrations and external calibrations. The recoveries of three organic Se species for spiked samples at 200 µg kg^{-1} levels all exceeded 90%. This method was suitable to accurately determine the three seleno-amino acids in eggs.

Supplementary Materials: The following supporting information can be downloaded at: https://www.mdpi.com/article/10.3390/separations9020021/s1, Figure S1: HPLC–ICP–MS chromatograms by Agilent SB-Aq column of egg sample extracted by different extraction procedures, Figure S2: HPLC–ICP–MS chromatograms by Agilent SB-Aq column of regular egg and Se-enriched egg, Table S1: Analytical performances of five Se species.

Author Contributions: Conceptualization, M.W.; methodology, Y.Z.; validation, Y.Z., M.Y. and J.Z.; investigation, Y.Z.; resources, M.W.; data curation, M.Y., J.Z. and T.W.; writing—original draft preparation, Y.Z.; writing—review and editing, M.Y., J.Z. and T.W.; supervision, M.W.; project administration, M.W. All authors have read and agreed to the published version of the manuscript.

Funding: This research was funded by the Ministry of Agriculture and Rural Affairs (MOARA) of the People's Republic of China (Special Fund for Agro-scientific Research), and Innovation Program of Chinese Academy of Agricultural Sciences.

Acknowledgments: The authors gratefully acknowledge the financial support of the Ministry of Agriculture and Rural Affairs (MOARA) of the People's Republic of China (Special Fund for Agro-scientific Research), and Innovation Program of Chinese Academy of Agricultural Sciences.

Conflicts of Interest: The authors declare no conflict of interest.

References

1. Ermakov, V.; Jovanović, L. Selenium deficiency as a consequence of human activity and its correction. *J. Geochem. Explor.* **2010**, *107*, 193–199. [CrossRef]
2. Bakırdere, S.; Volkan, M.; Ataman, O.Y. Selenium speciation in chicken breast samples from inorganic and organic selenium fed chickens using high performance liquid chromatography-inductively coupled plasma-mass spectrometry. *J. Food Compos. Anal.* **2018**, *71*, 28–35. [CrossRef]
3. Yang, G.Q.; Ge, K.Y.; Chen, J.S.; Chen, X.S. Selenium-related endemic diseases and the daily selenium requirement of humans. *World Rev. Nutr. Diet.* **1988**, *55*, 98–152. [CrossRef]
4. Miranda, J.M.; Anton, X.; Redondo-Valbuena, C.; Roca-Saavedra, P.; Rodriguez, J.A.; Lamas, A.; Franco, C.M.; Cepeda, A. Egg and egg-derived foods: Effects on human health and use as functional foods. *Nutrients* **2015**, *7*, 706–729. [CrossRef]
5. Kralik, G.; Kralik, Z. Poultry products enriched with nutricines have beneficial effects on human health. *Med. Glas. Off. Publ. Med. Assoc. Zenica-Doboj Canton Bosnia Herzeg.* **2017**, *14*, 1–7. [CrossRef]
6. Fisinin, V.I.; Papazyan, T.T.; Surai, P.F. Producing specialist poultry products to meet human nutrition requirements: Selenium enriched eggs. *World Poultry Sci. J.* **2008**, *64*, 85–97. [CrossRef]
7. Zhang, K.; Guo, X.; Zhao, Q.; Han, Y.; Zhan, T.; Li, Y.; Tang, C.; Zhang, J. Development and application of a HPLC-ICP-MS method to determine selenium speciation in muscle of pigs treated with different selenium supplements. *Food Chem.* **2020**, *302*, 125371. [CrossRef] [PubMed]
8. Tsopelas, F.; Tsantili-Kakoulidou, A.; Ochsenkühn-Petropoulou, M. Exploring the elution mechanism of selenium species on liquid chromatography. *J. Sep. Sci.* **2011**, *34*, 376–384. [CrossRef] [PubMed]
9. Jagtap, R.; Maher, W. Determination of selenium species in biota with an emphasis on animal tissues by HPLC-ICP-MS. *Microchem. J.* **2016**, *124*, 422–529. [CrossRef]
10. Ieggli, C.V.S.; Bohrer, D.; Noremberg, S.; do Nascimento, P.C.; de Carvalho, L.M.; Vieira, S.L.; Reis, R.N. Surfactant/oil/water system for the determination of selenium in eggs by graphite furnace atomic absorption spectrometry. *Spectrochim. Acta Part B-At. Spectrosc.* **2009**, *64*, 605–609. [CrossRef]
11. Lv, L.; Li, L.A.; Zhang, R.B.; Deng, Z.C.; Du, G.M.; Jin, T.M.; Yu, X.X.; Zhang, W.; Jiao, X.L. Effects of Dietary Supplementation of Nano-Selenium on the Egg Selenium Content and Egg Production Rate of North China Hens. *Nanosci. Nanotechnol. Lett.* **2018**, *10*, 1567–1571. [CrossRef]
12. Türker, A.R.; Erol, E. Optimization of selenium determination in chicken's meat and eggs by the hydride-generation atomic absorption spectrometry method. *Int. J. Food Sci. Nutr.* **2009**, *60*, 40–50. [CrossRef]
13. Lipiec, E.; Siara, G.; Bierla, K.; Ouerdane, L.; Szpunar, J. Determination of selenomethionine, selenocysteine, and inorganic selenium in eggs by HPLC–inductively coupled plasma mass spectrometry. *Anal. Bioanal. Chem.* **2010**, *397*, 731–741. [CrossRef] [PubMed]

14. Sun, H.; Feng, B. Speciation of Organic and Inorganic Selenium in Selenium-enriched Eggs by Hydride Generation Atomic Fluorescence Spectrometry. *Food Anal. Methods* **2011**, *4*, 240–244. [CrossRef]
15. Nagy, G.; Benko, I.; Kiraly, G.; Voros, O.; Tanczos, B.; Sztrik, A.; Takács, T.; Pocsi, I.; Prokisch, J.; Banfalvi, G. Cellular and nephrotoxicity of selenium species. *J. Trace Elem. Med. Biol.* **2015**, *30*, 160–170. [CrossRef]
16. Vicente-Zurdo, D.; Gomez-Gomez, B.; Perez-Corona, M.T.; Madrid, Y. Impact of fish growing conditions and cooking methods on selenium species in swordfish and salmon fillets. *J. Food Compos. Anal.* **2019**, *83*, 103275. [CrossRef]
17. Tie, M.; Li, B.; Sun, T.; Guan, W.; Liang, Y.; Li, H. HPLC-ICP-MS speciation of selenium in Se-cultivated Flammulina velutipes. *Arab. J. Chem.* **2020**, *13*, 416–422. [CrossRef]
18. Gao, H.H.; Chen, M.X.; Hu, X.Q.; Chai, S.S.; Qin, M.L.; Cao, Z.Y. Separation of selenium species and their sensitive determination in rice samples by ion-pairing reversed-phase liquid chromatography with inductively coupled plasma tandem mass spectrometry. *J. Sep. Sci.* **2018**, *41*, 432–439. [CrossRef]
19. Montes-Bayón, M.; Molet, M.J.D.; González, E.B.; Sanz-Medel, A. Evaluation of different sample extraction strategies for selenium determination in selenium-enriched plants (Allium sativum and Brassica juncea) and Se speciation by HPLC-ICP-MS. *Talanta* **2006**, *68*, 1287–1293. [CrossRef]
20. Yang, L.; Zhou, X.; Deng, Y.; Gong, D.; Luo, H.; Zhu, P. Dissipation behavior, residue distribution, and dietary risk assessment of fluopimomide and dimethomorph in taro using HPLC-MS/MS. *Environ. Sci. Pollut. Res.* **2021**, *28*, 43956–43969. [CrossRef]
21. Siwek, M.; Galunsky, B.; Niemeyer, B. Isolation of selenium organic species from antarctic krill after enzymatic hydrolysis. *Anal. Bioanal. Chem.* **2005**, *381*, 737–741. [CrossRef] [PubMed]
22. Zhang, Q.; Yang, G. Selenium speciation in bay scallops by high performance liquid chromatography separation and inductively coupled plasma mass spectrometry detection after complete enzymatic extraction. *J. Chromatogr. A* **2014**, *1325*, 83–91. [CrossRef] [PubMed]
23. Oliveira, A.F.; Landero, J.; Kubachka, K.; Nogueira, A.R.A.; Zanetti, M.A.; Caruso, J. Development and application of a selenium speciation method in cattle feed and beef samples using HPLC-ICP-MS: Evaluating the selenium metabolic process in cattle. *J. Anal. At. Spectrom.* **2016**, *31*, 1034–1040. [CrossRef]
24. Kurek, E.; Michalska-Kacymirow, M.; Konopka, A.; Kociuczuk, O.; Bulska, E. Searching for Low Molecular Weight Seleno-Compounds in Sprouts by Mass Spectrometry. *Molecules* **2020**, *25*, 2870. [CrossRef] [PubMed]
25. Schmidt, L.; Bizzi, C.A.; Duarte, F.A.; Muller, E.I.; Krupp, E.; Feldmann, J.; Flores, E.M.M. Evaluation of Hg species after culinary treatments of fish. *Food Control* **2015**, *47*, 413–419. [CrossRef]

Article

Quantitative Analysis of Anthocyanins in Grapes by UPLC-Q-TOF MS Combined with QAMS

Xue Li [1,2,3], Wei Wang [1], Suling Sun [1], Junhong Wang [1], Jiahong Zhu [1,2,3], Feng Liang [4], Yu Zhang [1,2,3,*] and Guixian Hu [1,*]

1. Institute of Agro-Product Safety and Nutrition, Zhejiang Academy of Agricultural Sciences, Hangzhou 310021, China; hei.semeng@163.com (X.L.); wangwei5228345@126.com (W.W.); sunsuling123@126.com (S.S.); wangjunhong@zaas.ac.cn (J.W.); zjnky2011@126.com (J.Z.)
2. Key Laboratory of Information Traceability for Agricultural Products, Ministry of Agriculture and Rural Affairs of China, Hangzhou 310021, China
3. Zhejiang Provincial Key Laboratory of Food Safety, Hangzhou 310021, China
4. College of Food Science and Technology, Bohai University, Jinzhou 121013, China; liangfeng1996@126.com
* Correspondence: zhyu7711@163.com (Y.Z.); hugx_shiny@163.com (G.H.); Tel.: +86-571-8641-7319 (G.H.)

Abstract: A method for quantifying the anthocyanins in grapes was firstly developed by ultra-high performance liquid chromatography-quadrupole-time of flight mass spectrometry (UPLC-Q-TOFMS) combined with quantitative analysis of multi-components by single marker (QAMS). A total of 10 main anthocyanins were analyzed by using peonidin 3-O-glucoside as the reference standard. The accuracy of this method was evaluated by an established and validated external standard quantification method with 10 reference compounds. The standard method difference (SMDs) of the quantification results between QAMS and the external standard method wasless than 15%. Furthermore, the QAMS method was used to analyzefour batches of grapes and the data was compared with those obtained using the external standard method. No significant difference wasobtained in the results obtained by both methods. These results indicated that the QAMS method could accurately determine the anthocyanins in grapes. This method can provide a basis to address the absence of reference standards for analyzing anthocyanins in other foods.

Keywords: anthocyanins; grapes; UPLC-Q-TOFMS; QAMS

Citation: Li, X.; Wang, W.; Sun, S.; Wang, J.; Zhu, J.; Liang, F.; Zhang, Y.; Hu, G. Quantitative Analysis of Anthocyanins in Grapes by UPLC-Q-TOF MS Combined with QAMS. *Separations* **2022**, *9*, 140. https://doi.org/10.3390/separations9060140

Academic Editor: Alena Kubatova

Received: 29 April 2022
Accepted: 25 May 2022
Published: 2 June 2022

Publisher's Note: MDPI stays neutral with regard to jurisdictional claims in published maps and institutional affiliations.

Copyright: © 2022 by the authors. Licensee MDPI, Basel, Switzerland. This article is an open access article distributed under the terms and conditions of the Creative Commons Attribution (CC BY) license (https://creativecommons.org/licenses/by/4.0/).

1. Introduction

Grapes are recognized as one of the most important commercial fruits worldwide [1]. Besides having adelightful flavor, they possess abundant nutrients and bioactive compounds [2]. In particular, a high content of anthocyanins, reaching up to 2300 mg/kg (fresh weight), isfound in red grapes [3]. Anthocyanins belong to water-soluble flavonoid-type polyphenols, which mainly exist in vacuoles of grape skin cells as free and acylated 3-O-glycosides derivatives [4]. Mostanthocyanin aglycones are based on anthocyanidins including cyanidin, delphinidin, petunidin, peonidin, pelargonidin, and malvidin, which share a 2-phenylbenzopyrilium skeleton hydroxylated in the 4′, 5′, and 7 positions [5]. The presence of anthocyanins is responsible for the sensory attributes of grapes, including aroma, taste, mouthfeel, and color [4]. Apart from their organoleptic properties, anthocyanins also have a variety of unique biological features that can promote human health. Some studies have proved that anthocyanins exhibited antioxidant [6], anti-tumor [7], anti-inflammatory [8], anti-diabetic [9], anti-obesity [10], anti-cardiovascular [11], and neuroprotective [12] properties. Accordingly, these compounds have aroused appreciable attention and have been the focus of many studies [2,13]. Therefore, a reliable method to quantify multiple bioactive anthocyanins in grapes is essential for their comprehensive quality control and improved utilization. However, despite the quantitative analyses of anthocyanins in grapes thathavebeen reported [14–17], most of these studies used the area normalization method to determine the chemical composition, which is a semi-quantitative procedure based on the chromatographic separation of analytes and a consistent detector response. The more commonly used external standard methods need standards for each chemical. It may be difficult to obtain due to a limited number and the high cost of commercially available anthocyanin standards. Considering these reasons, an accurate and sensitive method is necessary to solve the major bottleneck of reference compounds absence in the quantitative analysis of anthocyanins in grapes.

In the absence of reference standards, the quantitative analysis of complex components in foods is difficult. To tackle this urgency, a quantification method, quantitative analysis of multi-components by single marker (QAMS), has been first proposed in 2006 [18]. QAMS method is a relatively mature method to simultaneously detect the contents of multi-components in the sample, in which only one reference standard would be needed [19]. In QAMS method, each compound was quantified directly or calculated by a relative correction factor (RCF). QAMS can not only improve environmental friendliness, but decrease the operational complexity and experiment cost. This method has been widely used to analyze the components of Chinese herbal medicines [20,21], metabolites of aflatoxin [22], walnut leaves [23], and green tea extracts [24], and has been adopted by some national pharmacopoeia such as China, United States, and Europe [22]. Usually, QAMS was combined with high-performance liquid chromatography (HPLC) coupled with ultraviolet detection (UV) and/or mass spectrometry (MS) [24,25]. HPLC-UV has been proven to be an effective method for the multi-component analysis in plant-derived foods based on its good separation ability. However, its application can be limited because of low sensitivity, complex matrix interference, and the characterization of compounds only by the relative retention time [23]. Comparatively, HPLC-MS has advantages of high accuracy, sensitivity, and high separation efficiency to demonstrate trace detection ability [22], and provides molecular weight, characteristic fragment ions and retention time to ensure compound specificity [19]. Therefore, QAMS combined with HPLC-MS could be a high-effective method to quantify anthocyanins in grapes.

Due to complex anthocyanins existing in grapes, it is challenging to accurately and comprehensively analyze the anthocyanins of grapes. In this study, UPLC-Q-TOF MS was used to qualitatively screen the target anthocyanins in grapes through the molecular mass and retention time. The RCFs of QAMS were then determined by different calculation formulas and reference standards. Moreover, QAMS was used to carry out the quantitative analysis and the results were compared with those obtained by the external standard method. Finally, a sensitive and reliable QAMS coupled with the UPLC-Q-TOFMS technique was developed to quantify 10 main anthocyanins in grapes. To the best of our knowledge, using UPLC-Q-TOFMS coupled with the QAMS method for analysis of the anthocyanins in grapes has not been reported yet.

2. Materials and Methods

2.1. Chemicals and Reagents

Peonidin 3-O-glucoside chloride (Pn-G), malvidin 3-O-glucoside chloride (Mv-G), delphinidin 3-O-glucoside chloride (Dp-G), cyanidin 3-O-glucoside chloride (Cy-G), pelargonidin 3-O-glucoside chloride (Pe-G), petunidin 3-O-glucoside chloride (Pt-G), malvidin-3,5-O-diglucoside chloride (Mv-DG), peonidin-3,5-O-diglucoside chloride (Pn-DG), pelargonidin-3,5-O-diglucoside chloride (Pe-DG), and cyanidin-3,5-O-diglucoside chloride (Cy-DG) were purchased from Aladdin Reagent Co., Ltd. (Shanghai, China). The purities of the 10 reference standards were over 98% and their structures were shown in Figure 1. Acetonitrile (HPLC grade) was purchased from Merck (Darmstadt, Germany). Methanol (analytical grade) was obtained from Lingfeng Chemical Reagent Co., Ltd. (Shanghai, China). Ultrapure water (18 MΩ cm) was prepared using a purification system from Fulham Technology Co., Ltd. (Qingdao, China).

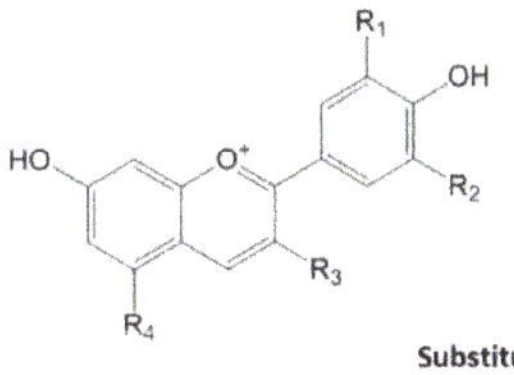

Anthocyanins	Substitutions			
	R_1	R_2	R_3	R_4
Peonidin 3-O-glucoside	OCH$_3$	H	Glc	OH
Malvidin 3-O-glucoside	OCH$_3$	OCH$_3$	Glc	OH
Delphinidin 3-O-glucoside	OH	OH	Glc	OH
Cyanidin 3-O-glucoside	OH	H	Glc	OH
Pelargonidin 3-O-glucoside	H	H	Glc	OH
Petunidin 3-O-glucoside	OCH$_3$	OH	Glc	OH
Malvidin-3,5-O-diglucoside	OCH$_3$	OCH$_3$	Glc	Glc
Peonidin-3,5-O-diglucoside	OCH$_3$	H	Glc	Glc
Pelargonidin-3,5-O-diglucoside	H	H	Glc	Glc
Cyanidin-3,5-O-diglucoside	OH	H	Glc	Glc

Figure 1. Structures of 10 anthocyanins (Glc: glucosyl).

Each solid analyte was weighed accurately and dissolved in methanol to prepare the stock solution (1 mg/mL). The mixed working solution was prepared by further dilution using 50% aqueous methanol daily for the optimization of extraction conditions. All the solutions were stored at 4 °C before analysis.

The samples named as "Meiselan", "Xila", "Chixiazhu1", and "Chixiazhu2" were obtained from Yantai City, Shandong Province of China.

2.2. Sample Preparation

The grape samples were prepared by a literature procedure [26]. Grape samples were pre-frozen with liquid nitrogen, and then ground to powder by a mortar and pestle. The accurately weighed grape powder (0.5 g) was dark extracted with 10 mL of 2% formic acid-methanol (v/v) in an ultrasonic bath at room temperature for 10 min. The extracts were on a shaker at 25 °C, 140 rpm/min, and centrifuged at 8000× g rpm for 10 min. The process was then repeated two times. Then the extracts were combined and evaporated on a rotary evaporator at 35 °C until dry. The residue was redissolved in 5 mL of methanol. The samples were filtered through a 0.22 μm millipore filter before analysis.

2.3. The Instrument Conditions

Waters Acquity UPLC system (Waters, Milford, MA, USA) was used toperform-Chromatographic analysis of the target anthocyanins. A Waters Acquity UPLC BEH C18 column (2.1 mm × 100 mm, 1.7 μm) was used in this study. The column temperature was set at 35 °C. The injection volume was 5 μL. The mobile phases were comprised of 1% formic acid aqueous solution (solvent A) and acetonitrile (solvent B) at a flow rate of 0.3 mL/min. An elution gradient was performed according to the following conditions: 0–1 min, 95.0–90.0% A; 1–6 min, 90.0–90.0% A, 6–9 min, 90.0–75.0% A; 9–11 min, 75.0–5.0% A; 11–13 min, 5.0–5.0% A; 13–13.5 min, 5.0–95.0% A; and 13.5–18 min, 95.0–95.0% A.

Mass spectrometry was performed on a Waters Q-TOFSynapt G2S high definition mass spectrometer (Waters, MA, USA). The quantification of the components was achieved on anelectrospray ionization source (ESI) in positive mode. The ion source working parameters were set as follows: source temperature, 80 °C; desolvation gas temperature, 300 °C; the flow rates of cone and desolvation gas, 50 L/h and 600 L/h, respectively; the voltages of capillary, cone and extraction cone, 3.0 kV, 35 V and 5.0 V, respectively. Full-scan mass range was 100–1000 Da.

2.4. Quantitative Analysis of Anthocyanins

In the present study, the anthocyanins were quantified by the QAMS method. The appropriate calculation method and reference analyte were selected to establish the relative correction factor (RCF) of each substance. RCF was applied to calculate the contents of the components. The external standard method (ESM) was used to verify the results of QAMS. The result with a lower standard method difference (SMD) compared with those from ESM was selected.

In the current study, RCF was calculated by the multipoint method (MP) and slope method (SP), respectively [23]. The formulas of RCF were as follow:

$$\text{RCF} = \frac{1}{n} \times \sum_{i=1}^{n} \frac{A_{si}}{C_{si}} \times \frac{C_{xi}}{A_{xi}} \quad (1)$$

$$\text{RCF} = \frac{K_s}{K_x} \quad (2)$$

Formula (1) was used to calculate the RCF via the MP method. Where As and Ax represented the peak areas of the reference analyte and analytes, respectively; Cs and Cx were the concentrations of the reference analyte and analytes, respectively; n represented the concentration numbers of the reference analyte and analytes; i represented the sum variables. Formula (2) was used to calculate the RCF the via SP method. Ks and Kx were the slopes of the standard curves.

$$C_x = \text{RCF} \times \frac{A_x \times C_s}{A_s} \quad (3)$$

Cs was calculated via ESM. Cx was determined based on Formula (3).

$$\text{SMD}(\%) = \frac{|C_{ESM} - C_{QAMS}|}{C_{ESM}} \times 100\% \quad (4)$$

SMD was calculated based on Formula (4), which could be used to indicate the difference between QAMS and ESM, and to verify the accuracy of QAMS. C_{ESM} and C_{QAMS} represented the concentrations of analytes calculated by ESM and QAMS methods, respectively.

2.5. Data Analysis

In this study, the data were represented by the mean and relative standard deviation (RSD) of three repeated experiments. GraphPad prism(v.5, GraphPad Software®, San Diego, CA, USA) was used for statistical analysis. The linear regression was performed using SPSS (v. 18.0, IBM®, Chicago, IL, USA). Statistical significance was set at the 95% confidence level ($p < 0.05$).

3. Results and Discussion

3.1. Optimization of the Instrumental Conditions

It is a challenging task to make chromatographic separation of anthocyanins due to their high diversity and structural similarity. However, MS detection allows simultaneous determination of chromatographically unresolved compounds by extraction of different m/z signals [25]. In this study, several parameters were optimized, including different concentrations of formic acid aqueous solution (water with formic acid—0%, 0.5%, and 1%), the organic phases (methanol and acetonitrile), gradient elution program, and different chromatographic columns (Waters ACQUITY UPLC BEH C18 (100 mm × 2.1 mm, 1.7 μm), Waters ACQUITY UPLC RP18 (100 mm × 2.1 mm, 1.7 μm)). The results showed that the best separation degree and peak shape were obtained using Waters ACQUITY UPLC BEH C18 (100 mm × 2.1 mm, 1.7 μm) column and a gradient elution program with acetonitrile and water containing 1% formic acid. Typical extract ion chromatograms of the anthocyanin standards and grape extract were shown in Figure 2.

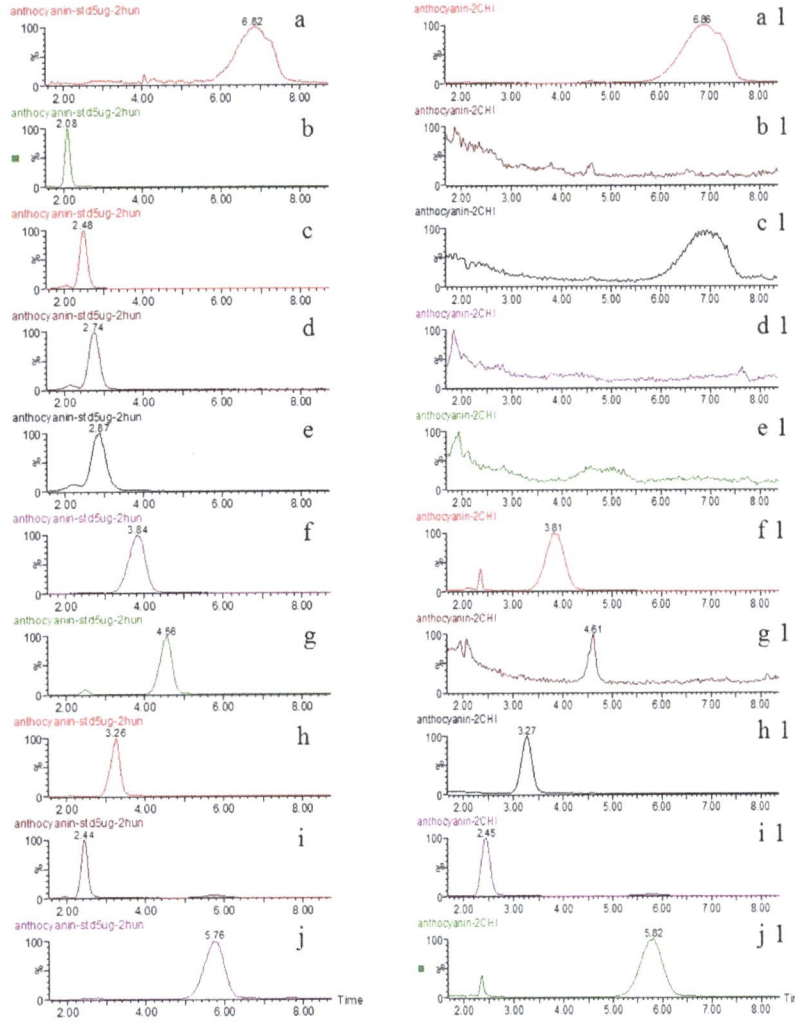

Figure 2. Typical extract ion chromatograms of mixed reference standards (**a–j**) and the grape extract (**a1–j1**) ((**a,a1**), Mv-G; (**b,b1**) Cy-DG; (**c,c1**), Pe-DG; (**d,d1**), Pn-DG; (**e,e1**), Mv-DG; (**f,f1**), Pt-G; (**g,g1**), Pe-G; (**h,h1**) Cy-G; (**i,i1**), Dp-G; (**j,j1**), Pn-G).

3.2. Validation of the Method

The retention time and MS^1 mass were got under the optimized instrumental conditions. In order to verify the applicability of the method, the linear ranges, regression coefficients, the limit of detection (LOD), the limit of quantification (LOQ), precisions, and repeatability were performed. The linearity was established with the peak areas of six different concentrations for each anthocyanin. As shown in Table 1, good linear relationships were obtained with satisfactory correlation coefficients (R^2) greater than 0.99. The LOD and LOQ were calculated as the signal to noise ratios (S/N) of 3 and 10, respectively. The instrument precision was evaluated by the relative standard deviation (RSD) calculated for the intra-day and inter-day variations. Both RSD values of the intra-day (0.58–4.09%) and inter-day (0.61–4.23%) variations were within the acceptable range. The precision of this method was evaluated by repeatability. Six repeated samples from the same batch were measured by the developed method, and the results (RSD)were in the range of 1.22–2.43%. The above results indicated that the method was considered to be effective and reliable. Therefore, this method could be used for the quantitative analysis of anthocyanins in grapes.

Table 1. The results of method validation for the 10 analytes.

No.	Compound	Regression Equation	Linearity Range (µg/mL)	Correlation Coefficient (R^2)	LOD (µg/mL)	LOQ (µg/mL)	Precision		Repeatability
							Intra-Day RSD (%)	Inter-Day RSD (%)	RSD (%)
1	Pn-G	Y = 9558X + 1598.3	0.25–5.00	0.9992	0.05	0.20	1.33	0.79	1.22
2	Mv-G	Y = 7824.3X + 740.48	0.24–4.88	0.9999	0.05	0.20	2.62	1.65	1.76
3	Dp-G	Y = 2387X + 2139.5	0.50–9.90	0.9943	0.10	0.40	3.08	1.71	2.16
4	Cy-G	Y = 5336.4X + 3894.4	0.75–15.00	0.9988	0.15	0.50	3.99	0.61	2.22
5	Pe-G	Y = 8655.4X + 1850.4	0.25–4.99	0.9963	0.05	0.20	3.45	2.35	2.43
6	Pt-G	Y = 5021X + 2382.5	0.50–9.90	0.9949	0.10	0.40	4.00	2.28	1.59
7	Mv-DG	Y = 547.08X + 1619.8	1.75–34.95	0.9958	0.50	1.50	3.31	1.87	1.95
8	Pn-DG	Y = 3951.6X + 2697.6	0.37–7.42	0.9932	0.10	0.30	0.58	0.80	2.01
9	Pe-DG	Y = 9817X + 6488.5	0.75–7.5	0.9951	0.15	0.50	4.45	4.23	2.15
10	Cy-DG	Y = 2130.2X + 3240.5	1.25–12.5	0.9922	0.30	1.00	4.09	3.72	2.05

3.3. QAMS Method Development

At present, the QAMS approach coupled with HPLC-UV or HPLC-QqQ-MS was widely used in the quantitative analysis [19,24]. However, these methods could not be used for the qualitative and quantitative analysis of complex matrix samples simultaneously. Therefore, for better authentication and convenience forquantitative determination of compounds, the position of target compounds needs to be further corrected by the relative retention time of these compounds with reference substancesin different columns and instruments [20]. UPLC-Q-TOF MS is an instrument based on the quadrupole-time of flight technique. The method based on the instrument could not only reduce the matrix interference, but also quantify the analytes with low and high contentssimultaneously and preliminarily identify the analytes by the mass spectrometric data without reference compounds. Thus, it could be a powerful tool for the qualitative and quantitative analysis of complex matrix samples at the same time. In the present research, UPLC-Q-TOF MS was combined with QAMS to determine the anthocyanins in grapes. The obtained chromatographic peaks were identified by comparing the retention times, molecular weight, and fragment ions with chemical reference substances (Table 2).

Table 2. Parameters for the 10 analytes in MS mode.

No.	Compound	PrecursorIon (*m/z*)	Fragment Ion (*m/z*)	Rt (min)
1	Pn-G	463.2524	301	5.76
2	Mv-G	493.2640	331	6.82
3	Dp-G	465.1743	303	2.44
4	Cy-G	449.2361	287	3.26
5	Pe-G	433.2406	271	4.56
6	Pt-G	479.2479	317	3.84
7	Mv-DG	655.2778	331	2.87
8	Pn-DG	625.2689	301	2.74
9	Pe-DG	595.2590	271	2.48
10	Cy-DG	611.2512	287	2.08

QAMS method was designed according to the principle of linear relationship between the component amount and the detector response within a certain range [22]. Forthe QAMS method, RCFs between the components werea critical parameter because its ruggedness and robustness heavily influence the accuracy of the QAMS method [21]. The value of RCF was affected by many factors. Usually, the factors including the instrument, mobile phase, pH, flow rate, chromatographic column, and the column temperature, were the most often considered [22,24], while the previous results indicated that these factors had little effect on the RCF. Up to now, the RCF of the QAMS method could be divided into two types based on their calculation methods. One was calculated by the ratio of slopes of the analytes (SP); the other was calculated using the average of several RCFs from the referring standard and the analyte detected under multiple concentration levels (MP) [21]. However, few studies have compared the results of the two methods. Moreover, the selection of a reference standard significantly affects the accuracy of RCF. However, very few studies have focused on the choice of a single marker. Usually, the reference substance was a cheap and readily available typical component and was used to determine the RCF between the analytes and the reference component. Anthocyanins in grapes are complex and diverse, while the structure of the reference analyte might influence the RCF. Therefore, we focused on the selection of reference analyte and the calculation method of RCF in this study.

The RCF values of analytes were calculated with different reference standards according to Formulas (1) and (2). The concentration contents of 10 anthocyanins in the test sample were determined by QAMS and ESM methods. In ESM, the contents of all the analytes were calculated based on linear regression equations listed in Table 1. In QAMS, the contents of the target analytes were calculated by Equation (3) with the different RCFs. To select the proper formula of RCF, the values of the SMDs were determined with the help of Equation (4) as the evaluation principle. The boxplots of the SMDs were shown in Figure 3, and the MP and SP methods were applied to calculatethe RCF values, respectively. According to the results, the SMDs calculated from the RCF values obtained by MP were all higher than those obtained by SP. It indicated that the quantification of the analytes with the RCF values calculated by SP was more accurate than that obtained by MP. It may be due to the fluctuations at different concentration levels. As described previously, SP was ultimately selected to calculate RCF values. On the other hand, the selection of the reference standard is very important for the QASM method. Figure 3b shows that Pn-G could be regarded as a reference standard for calculatingthe RCF values of the other compounds because the SMDs of all components with Pn-G as a reference standard were lower than those of the othersevenreference standards. In addition, Pn-Ghas the advantages of simple structure, low price, and easily obtained. Moreover, the SMDs of the other nineanthocyanins calculated with Pn-G as reference standard were between 0.15% and 13.66%. Therefore, it should be noted that the SMDs of the analytes with different structures calculated by using the same reference standard wasdifferent. If Mv-G, Dp-G, Cy-G, Pt-G, Mv-DG, Pn-DG, and Cy-DG were used as reference standards, the SMDs of the 10 anthocyanins were 4.93–27.94%, 4.23–23.05%, 3.14–30.12%, 0.61–22.12%, 1.60–23.37%, 6.21–40.07%, and 2.36–25.68%, respectively. The results indicated that the structure of reference standard and analytes could influence the values of RCF and the accuracy of the QAMS method. This result was consistent with that of the literature [21]. Therefore, the established QAMS method was more appropriate for the quantitative analysis of multiple compounds with similar structures in foods. Finally, to compare the difference between ESM and QAMS, the t-test was used for statistical analysis, and the p values were all greater than 0.05. The results are shown in Table 3. Therefore, the QAMS, which used Pn-G as the reference standard and adopted the SP method to calculate the RCF values, can simultaneously determine the 10 anthocyaninsinstead of the ESM.

Table 3. The contents of the 10 anthocyanins by ESM and QAMS (μg/mL).

Compounds	Quantitative Method	Mean	RSD%	p
Pn-G	ESM	2.07	2.50	/
Mv-G	ESM	1.96	1.79	0.63
	QAMS	1.98	3.77	
Dp-G	ESM	4.94	0.54	0.63
	QAMS	4.96	1.04	
Cy-G	ESM	6.39	1.18	0.56
	QAMS	6.35	1.40	

Table 3. Cont.

Compounds	Quantitative Method	Mean	RSD%	p
Pe-G	ESM	1.92	4.90	0.56
	QAMS	1.96	4.15	
Pt-G	ESM	4.19	3.41	0.42
	QAMS	4.09	3.07	
Mv-DG	ESM	18.64	1.31	0.08
	QAMS	18.28	0.66	
Pn-DG	ESM	4.07	0.74	0.06
	QAMS	3.86	3.43	
Pe-DG	ESM	5.60	1.19	0.78
	QAMS	5.62	1.07	
Cy-DG	ESM	10.44	0.97	0.15
	QAMS	10.20	2.14	

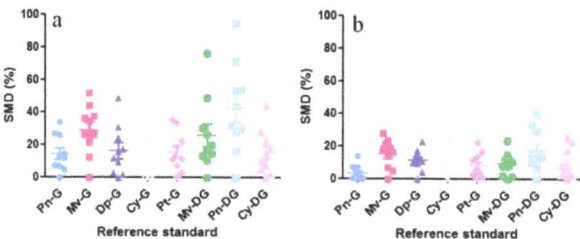

Figure 3. Boxplots of the differences in quantification results between ESM and QAMS: (**a**) the SMDs of all anthocyanins with different anthocyanins as reference standards, where F is calculated by MP; (**b**) the SMDs of all anthocyanins with different anthocyanins as reference standards, where F is calculated by SP.

3.4. Application of Proposed Method to Grape Samples

To validate the feasibility of QAMS to determine multi-compounds in grapes, the 10 anthocyanins contents were determined by ESM and QAMS in four batches, respectively. Six anthocyanins were detected in these grapes including Pn-G, Dp-G, Cy-G, Pe-G, Pt-G and Mv-G and the contents of Pe-G were all below the limit of quantitation. The linear regression model was built between the two variables to measure the deviation between QAMS and ESM (Table 4). The independent and dependent variables were determined by QAMS and ESM, respectively. In the regression model, all statistically significant coefficients were 0.000, and the R^2 values of Dp-G, Cy-G, Pt-G, and Mv-G were 1.000, 1.000, 0.999, and 0.999, respectively. The results indicated that there was a good statistically significant correlation between the two variables, and no significant difference in the contents obtained by QAMS and ESM. Therefore, QAMS could be used to determine anthocyanins in grapes.

Table 4. The summary of the linear regression model and ANOVA.

	Dp-G	Cy-G	Pt-G	Mv-G
R	1.000	1.000	1.000	1.000
R^2	1.000	0.999	0.999	0.999
F	7326.6	2435.9	2165.0	2085.4
Sig.	0.000	0.000	0.000	0.000

4. Conclusions

In this study, a method for quantifying anthocyanins in grapes was established by UPLC-Q-TOF MS combined with QAMS, and the different calculation formulas of RCF and the reference standards were compared. SP was ultimately selected to calculate RCF values and Pn-G was chosen as a reference standard to determine the 10 compounds' contents. The SMDs between QAMS and ESM were below 15%. No significant deviations of anthocyanin contents between QAMS and ESM were obtained in the four batches of grapes. The results showed that the established QAMS method could replace the ESM method under the condition of a lack of a reference standard. The established QAMS method displayed the advantages of simplicity, accuracy, and low price. It may provide new ideas for the quantitative study of anthocyanin in other foods like colored grain, fruits, and vegetables, etc.

Author Contributions: X.L.: methodology, formal analysis, investigation, writing—original draft; W.W.: conceptualization, supervision; S.S.: validation; J.W.: methodology; J.Z.: methodology; F.L.: resources; Y.Z.: data curation, funding acquisition; G.H.: supervision, funding acquisition. All authors have read and agreed to the published version of the manuscript.

Funding: This work was supported by China Agriculture Research System (CARS-29), the Analysis and Measurement Foundation of Zhejiang Province (No. LGC21C200002) and the Key Research and Development Program of Zhejiang Province (No. 2021C02002).

Institutional Review Board Statement: Not applicable.

Informed Consent Statement: Not applicable.

Conflicts of Interest: The authors declare that they have no known competing financial interests or personal relationships that could have appeared to influence the work reported in this paper.

References

1. Zhao, X.; Zhang, S.-S.; Zhang, X.-K.; He, F.; Duan, C.-Q. An effective method for the semi-preparative isolation of high-purity anthocyanin monomers from grape pomace. *Food Chem.* **2020**, *310*, 125830. [CrossRef] [PubMed]
2. Sabra, A.; Netticadan, T.; Wijekoon, C. Grape bioactive molecules, and the potential health benefits in reducing the risk of heart diseases. *Food Chem. X* **2021**, *12*, 100149. [CrossRef] [PubMed]
3. Han, F.; Yang, P.; Wang, H.; Fernandes, I.; Mateus, N.; Liu, Y. Digestion and absorption of red grape and wine anthocyanins through the gastrointestinal tract. *Trends Food Sci. Technol.* **2019**, *83*, 211–224. [CrossRef]
4. Crupi, P.; Palattella, D.; Corbo, F.; Clodoveo, M.L.; Masi, G.; Caputo, A.R.; Battista, F.; Tarricone, L. Effect of pre-harvest inactivated yeast treatment on the anthocyanin content and quality of table grapes. *Food Chem.* **2021**, *337*, 128006. [CrossRef]
5. Lima, Á.S.; de Oliveira, B.S.; Shabudin, S.V.; Almeida, M.; Freire, M.G.; Bica, K. Purification of anthocyanins from grape pomace by centrifugal partition chromatography. *J. Mol. Liq.* **2021**, *326*, 115324. [CrossRef]
6. Kharadze, M.; Japaridze, I.; Kalandia, A.; Vanidze, M. Anthocyanins and antioxidant activity of red wines made from endemic grape varieties. *Ann. Agrar. Sci.* **2018**, *16*, 181–184. [CrossRef]
7. Zhou, L.; Wang, H.; Yi, J.; Yang, B.; Li, M.; He, D.; Yang, W.; Zhang, Y.; Ni, H. Anti-tumor properties of anthocyanins from Lonicera caerulea 'Beilei' fruit on human hepatocellular carcinoma: In vitro and in vivo study. *Biomed. Pharmacother.* **2018**, *104*, 520–529. [CrossRef]
8. Decendit, A.; Mamani-Matsuda, M.; Aumont, V.; Waffo-Teguo, P.; Moynet, D.; Boniface, K.; Richard, E.; Krisa, S.; Rambert, J.; Mérillon, J.-M.; et al. Malvidin-3-O-β glucoside, major grape anthocyanin, inhibits human macrophage-derived inflammatory mediators and decreases clinical scores in arthritic rats. *Biochem. Pharmacol.* **2013**, *86*, 1461–1467. [CrossRef]
9. Gowd, V.; Jia, Z.; Chen, W. Anthocyanins as promising molecules and dietary bioactive components against diabetes—A review of recent advances. *Trends Food Sci. Technol.* **2017**, *68*, 1–13. [CrossRef]
10. Xie, L.; Su, H.; Sun, C.; Zheng, X.; Chen, W. Recent advances in understanding the anti-obesity activity of anthocyanins and their biosynthesis in microorganisms. *Trends Food Sci. Technol.* **2018**, *72*, 13–24. [CrossRef]
11. de Pascual-Teresa, S. Molecular mechanisms involved in the cardiovascular and neuroprotective effects of anthocyanins. *Arch. Biochem. Biophys.* **2014**, *559*, 68–74. [CrossRef] [PubMed]
12. Suresh, S.; Begum, R.F.; Singh, A.S.; Chitra, V. Anthocyanin as a therapeutic in Alzheimer's disease: A systematic review of preclinical evidences. *Ageing Res. Rev.* **2022**, *76*, 101595. [CrossRef] [PubMed]
13. de Oliveira Filho, J.G.; Braga, A.R.C.; de Oliveira, B.R.; Gomes, F.P.; Moreira, V.L.; Pereira, V.A.C.; Egea, M.B. The potential of anthocyanins in smart, active, and bioactive eco-friendly polymer-based films: A review. *Food Res. Int.* **2021**, *142*, 110202. [CrossRef]
14. Xie, S.; Liu, Y.; Chen, H.; Zhang, Z.; Ge, M. Anthocyanin degradation and the underlying molecular mechanism in a red-fleshed grape variety. *LWT* **2021**, *151*, 112198. [CrossRef]
15. Ju, Y.; Yang, L.; Yue, X.; Li, Y.; He, R.; Deng, S.; Yang, X.; Fang, Y. Anthocyanin profiles and color properties of red wines made from Vitisdavidii and Vitis vinifera grapes. *Food Sci. Hum. Wellness* **2021**, *10*, 335–344. [CrossRef]
16. Chen, S.; Zhang, F.; Ning, J.; Liu, X.; Zhang, Z.; Yang, S. Predicting the anthocyanin content of wine grapes by NIR hyperspectral imaging. *Food Chem.* **2015**, *172*, 788–793. [CrossRef]

17. Pinelli, P.; Romani, A.; Fierini, E.; Agati, G. Prediction models for assessing anthocyanins in grape berries by fluorescence sensors: Dependence on cultivar, site and growing season. *Food Chem.* **2018**, *244*, 213–223. [CrossRef]
18. Fan, C.J. Application situation of multi-components quantitation by one marker new method for quality evaluation and control of Chinese herbal medicine. *Pharm. Clin. Chin. Mater. Med.* **2013**, *4*, 18–20.
19. Ning, Z.; Liu, Z.; Song, Z.; Zhao, S.; Dong, Y.; Zeng, H.; Shu, Y.; Lu, C.; Liu, Y.; Lu, A. A single marker choice strategy in simultaneous characterization and quantification of multiple components by rapid resolution liquid chromatography coupled with triple quadrupole tandem mass spectrometry (RRLC-QqQ-MS). *J. Pharm. Biomed. Anal.* **2016**, *124*, 174–188. [CrossRef]
20. Zhu, C.; Li, X.; Zhang, B.; Lin, Z. Quantitative analysis of multi-components by single marker—a rational method for the internal quality of Chinese herbal medicine. *Integr. Med. Res.* **2017**, *6*, 1–11. [CrossRef]
21. Wang, C.-Q.; Jia, X.-H.; Zhu, S.; Komatsu, K.; Wang, X.; Cai, S.-Q. A systematic study on the influencing parameters and improvement of quantitative analysis of multi-component with single marker method using notoginseng as research subject. *Talanta* **2015**, *134*, 587–595. [CrossRef] [PubMed]
22. Wang, X.; Zhao, Y.; Qi, X.; Zhao, T.; Wang, X.; Ma, F.; Zhang, L.; Zhang, Q.; Li, P. Quantitative analysis of metabolites in the aflatoxin biosynthesis pathway for early warning of aflatoxin contamination by UHPLC-HRMS combined with QAMS. *J. Hazard. Mater.* **2022**, *431*, 128531. [CrossRef] [PubMed]
23. Su, C.; Li, C.; Sun, K.; Li, W.; Liu, R. Quantitative analysis of bioactive components in walnut leaves by UHPLC-Q-Orbitrap HRMS combined with QAMS. *Food Chem.* **2020**, *331*, 127180. [CrossRef] [PubMed]
24. Li, D.-W.; Zhu, M.; Shao, Y.-D.; Shen, Z.; Weng, C.-C.; Yan, W.-D. Determination and quality evaluation of green tea extracts through qualitative and quantitative analysis of multi-components by single marker (QAMS). *Food Chem.* **2016**, *197*, 1112–1120. [CrossRef] [PubMed]
25. Stekolshchikova, E.; Turova, P.; Shpigun, O.; Rodin, I.; Stavrianidi, A. Application of quantitative analysis of multi-component system approach for determination of ginsenosides in different mass-spectrometric conditions. *J. Chromatogr. A* **2018**, *1574*, 82–90. [CrossRef]
26. He, J.J.; Liu, Y.X.; Pan, Q.H.; Cui, X.Y.; Duan, C.Q. Different Anthocyanin Profiles of the Skin and the Pulp of Yan73 (Muscat Hamburg × Alicante Bouschet) Grape Berries. *Molecules* **2010**, *15*, 1141–1153. [CrossRef]

MDPI
St. Alban-Anlage 66
4052 Basel
Switzerland
Tel. +41 61 683 77 34
Fax +41 61 302 89 18
www.mdpi.com

Separations Editorial Office
E-mail: separations@mdpi.com
www.mdpi.com/journal/separations

www.ingramcontent.com/pod-product-compliance
Lightning Source LLC
LaVergne TN
LVHW070602100526
838202LV00012B/539